Simplified Guide to Electronic Circuits, Test Procedures and Troubleshooting

Simplified Guide to Electronic Circuits, Test Procedures and Troubleshooting

Hugh Moon

(with illustrations by Kermit Martin)

Parker Publishing Company, Inc.
West Nyack, New York

PARKER PUBLISHING COMPANY, INC.
West Nyack, N.Y.

Library of Congress Cataloging in Publication Data

Moon, Hugh
Simplified guide to electronic circuits, test procedures and troubleshooting.

Includes index.
1. Radio. 2. Television. 3. Electronic circuits.
I. Title.
TK6550.M557 621.3841'2 75-4562
ISBN 0-13-810036-5

Printed in the United States of America

To my wife Deane—
whose help and understanding
made writing this book possible.

The Scope and Practical Value of This Book

This book offers a unique method of learning how to understand and work successfully with electronic circuits. By combining practical theory and the advantages of a laboratory manual in one book, the reader will quickly learn how a circuit functions. This will be reinforced by a detailed illustration of one of the more commonly used circuits. You will also be told where and how to use test instruments in maintaining and troubleshooting circuits.

Simplicity of theory and illustrations will be of particular value to technicians who have a knowledge of the basic principles and operation of components used in electronic circuits. However, anyone —even those without a great deal of technical knowledge—with a desire to learn more about the fascinating world of electronic circuitry, will acquire a clear understanding of this essential aspect of electronics through the helpful, detailed illustrations.

Information in this book is presented in the form of a logical, progressive series of facts, one step leading to another, and understanding each of these steps will lead to a far clearer, more productive mastery of troubleshooting techniques. In the beginning we will deal with the essential tools of the electronic technician: reliable test equipment. Guidelines and specific suggestions for the most effective way to use this equipment for time-saving repairs will be described in detail.

Each succeeding chapter describes an essential circuit, explaining and showing through illustrations how test equipment is applied to each circuit. There is also a chapter that will help to simplify the complex circuitry of television, including several unique and effective troubleshooting methods when servicing this equipment.

The book describes and illustrates basic electronic circuits which are frequently referred to by name. You will gradually advance to higher, more intricate levels of electronics, but very often will find the same basic circuit with only the name changed, even though the circuit may be a more elaborate configuration. For example, in Chapter 11 a transistor amplifier is described and illustrated, including how the output signal is the reverse of that at the input. In pulse-operated circuits where polarity of the pulse must be changed, the amplifier is simply called an inverter.

You will learn in Chapter 4 how the oscillator circuit operates, and where the many different types are used, along with methods and equipment necessary to test and troubleshoot these circuits. Chapter 12 explains the multivibrator, and its usage in counters and pulse-switching circuits. As you know, the modern television receiver of today often uses the multivibrator in the vertical and horizontal oscillator circuits. The automotive industry also announced recently the use of electronic ignition systems on many of the newer cars, a development that stems largely from the multivibrator circuit.

Despite the fact that new electronic circuit uses are regularly announced by the industry, the basic circuits shown in this book will nevertheless be your key toward understanding, servicing, and keeping pace with these newer developments.

Electronic technology is advancing at a terrific pace, but regardless of the electronic circuit used in a complex piece of equipment the *basic* circuit will come from some of those described in this book. For example, the heart of the computer is the clock generator which is a crystal-controlled oscillator. This is described in detail in Chapter 4, along with illustrations that show how to connect test instruments for troubleshooting this particular circuit.

Even instruments used in the medical profession that produce high frequency sound waves are basically the result of multivibrator circuits, described in Chapter 12. This chapter also includes illustrations on how to connect and use test instruments most effectively.

All technicians will find this a *useful* book and also a practical refresher course in many of the basic but often overlooked or forgotten principles in analyzing electronic circuitry. It will enrich your practical knowledge and enable you to make faster and more rewarding progress in the important field of electronics.

Hugh Moon

Table of Contents

Simplified Guide to Electronic Circuits, Test Procedures and Troubleshooting

1

Understanding the Importance of Test Equipment

TROUBLESHOOTING

As an electronic technician, you may be called upon to install new equipment, to maintain equipment that has been installed, or in case of failure of this equipment, to repair it. In order to perform this job efficiently, there are many test devices available to help you.

When installing new equipment and performing routine maintenance, you have as your aid such equipment as signal generators, both AF (audio frequency) and RF (radio frequency), voltmeters, ammeters, frequency meters, and output meters. In case of equipment failure you may use vacuum tube testers, vacuum tube voltmeters, signal tracers, oscilliscopes, and resistance-capacitance substitution boxes. There are many more devices to aid you in doing a more efficient and timesaving job.

A carpenter would not start building a house with only a tack hammer and a saw. An electronics technician should not start troubleshooting any type of electronic device with only a neon bulb and a screwdriver.

Modern electronic devices with printed circuits and modular construction using IC chips and operated from a pulse generator must be maintained by using accurate and dependable test equipment. With the discrete circuitry we could visually inspect it for possible trouble; but for today's IC chips and modular circuitry you must have a more accurate test instrument, like the oscilloscope, which will look inside the circuit for possible trouble.

THE PRINCIPAL PURPOSE FOR TEST EQUIPMENT

Accurate and dependable test equipment in the hands of a good electronics technician can reduce the down time of an electronic device.

For example: you are the engineer at a television station, and very suddenly the station goes off the air. Your knowledge and good test equipment could save the station money, because every minute the transmitter is down, the station is losing money.

Another example: you are in the TV repair business. It is Saturday morning and Mr. Jones brings to your shop his TV for repair. He wants it repaired while he waits. His wife is home with three children who are driving her crazy, and she needs that TV so the children can watch the cartoons and be quiet. Here your knowledge and good test equipment could make you a hero for the Jones family.

Good test equipment used and displayed in your shop makes a very good impression on your customers. They know they are getting the best, and they are going to tell their friends.

Almost anyone can repair electronic devices by replacing defective parts, after diagnosing defects through troubleshooting methods using the proper test equipment. Today's electronic technician is striving for *speed, accuracy,* and *results*. Without the proper equipment, his speed is retarded, his accuracy is questionable, and his results are few. "A technician is only as good as his equipment."

BEST WAYS TO USE AND OPERATE TEST EQUIPMENT

When you purchase a piece of test equipment, the first thing you should do is read the enclosed instructions. The manufacturer has designed the equipment to perform a certain function and has written instructions for you to show how it should perform. Familiarize yourself with all controls, checking the function of each and what relationship they may have with one another. Check the leads; make sure that you are familiar with any switch on the probe that could change the function of the instrument. Many vacuum tube voltmeters have a switch on the probe which changes the meter reading from AC to DC. Many oscilloscopes have different probes that must be changed when testing different circuits. Some instruments use a ventilating fan which has a filter on the outside. This should be cleaned often to avoid overheating. Keep the outside of your instruments clean of dust and finger marks. You will be surprised how much better you feel using them when they are clean, and how much more impressive they appear to any visitor who observes your work bench. Never set an instrument on top of another instrument, or on the edge of your work bench. Remember, this instrument belongs to you, and you will be the one who has to replace it when it comes crashing to the floor accidentally.

Until you are thoroughly familiar with your instrument, look twice before connecting it to a circuit. Be sure the controls are in the proper position for the function you are going to perform. Be sure the probe is correct for the polarity you are connected to. When you are using someone else's instruments, be as careful with them as you would be with your own. Think how you would feel if it belonged to you and someone else damaged it. Remember, your test equipment may be earning your livelihood, so treat it with respect.

HOW TO MAKE THE MOST EFFECTIVE USE OF TEST EQUIPMENT

The volt-ohm-milliammeter (VOM) is more often referred to as a multimeter, because it is really three meters combined in one case. A switch on the front selects the proper mode for which the meter can be used, combining the voltmeter, ohmmeter and ammeter, all with one meter movement.

Refer to Figure 1-1 (A). This shows the meter face. Notice that all three values are marked on separate scales. Usually the top scale is the ohms. The second scale is usually the DC volts and DC current scale. The bottom scale is for measuring AC voltage. This meter will not measure AC current; a special meter is required for that.

Figure 1-1 (B) shows the bottom half of the case with the selector switch in the center; the upper left knob selects AC or DC voltage. The knob at the upper right adjusts the meter to zero; when reading resistance, the bottom shows the leads and their polarity. When the selector switch is pointed to the top center, the meter will read ohms and is marked R × 1, indicating the scale is multiplied by one; read direct. The next position of the selector switch, going in a clockwise direction, reads R × 10; ohms scale is then multiplied by 10. The next reading, continuing clockwise, is R × 100; the ohms scale is then

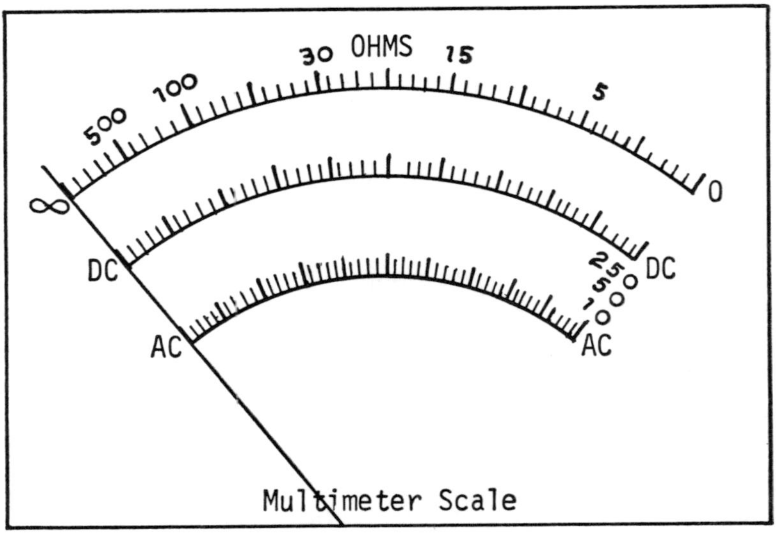

FIGURE 1-1(A): Multimeter.

multiplied by 100. When the selector switch is in the R × 1 range, the meter leads must be connected together, and the ohms adjust knob adjusted until the meter reads zero. The reason for this is that a low voltage battery is in series with the meter for this range. When the selector switch is in the R × 10 and R × 100 range, there is a higher voltage battery switched into the circuit, so the meter must be recalibrated.

Continuing in a clockwise direction, the meter now has been switched to read voltage. If the voltage is DC, then the AC-DC switch is turned to DC. Usually the numbers are between the DC and AC scales and are used for both voltages.

The DC scale is a different color to distinguish it from the AC. The first position of the selector knob shows 10v. The DC scale now reads from zero to 10 volts—the maximum that can be read on this scale. The next range is 250 volts. The meter scale now reads from zero to 250 volts. The next range is 500 volts. On some meters this range is read from zero to 50 volts and multiplied by 10. When the selector switch is placed on any range, this is the maximum voltage that can be read on this range.

To read AC voltage, the AC-DC switch is turned to AC, using the same numbers and multipliers as before, and rotating the switch to the

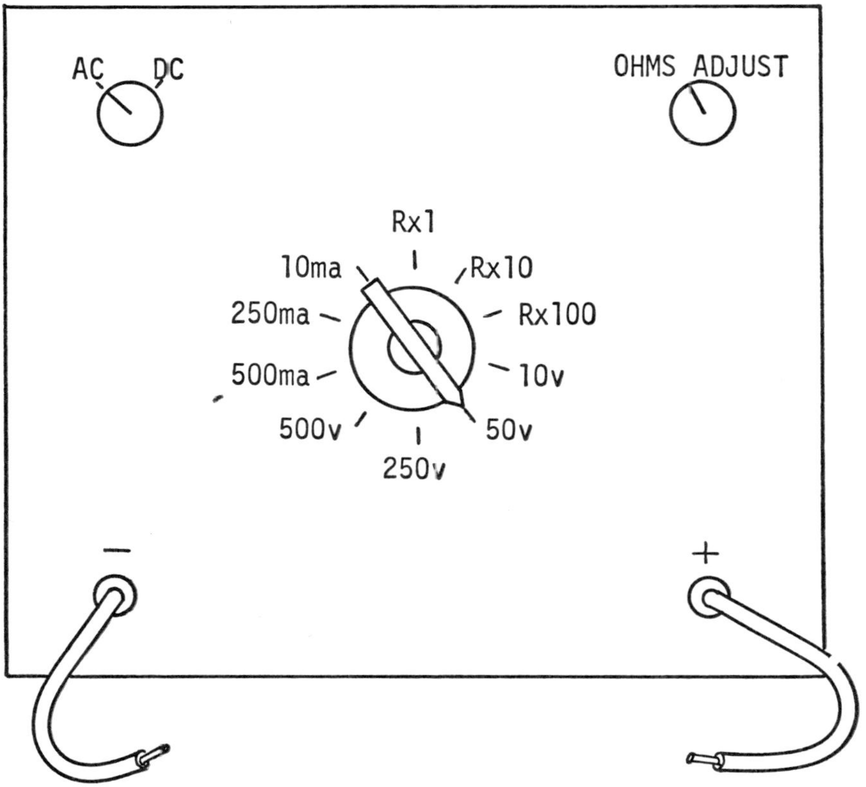

FIGURE 1-1(B): Multimeter controls.

next position clockwise. At this position the meter will read current, but the AC-DC switch must be switched back to DC; then turn the selector switch to 10 ma. Use the same scale that was used to read DC voltage. With the switch in this position it will read from zero to 10 ma. Next, the selector switch is rotated counterclockwise to 250 ma. The scale now reads from zero to 250 ma. Continuing counterclockwise, this scale is 500 ma. Use the numbers from zero to 50 ma, and multiply each number by 10. This scale will read from zero to 500 ma. The maximum that can be read when the switch is rotated is the last number on the scale that is chosen.

Some meters have an extra jack on the front of the case which is marked 10 A. This extends the current range to 10 amperes by placing the positive lead in this jack, and placing the selector switch to 10 ma. The scale then will read from zero to 10 A.

There are many different sizes in the multimeter cases, also different kinds of meter foundations having different sensitivity. This may vary the values on the selector switch, also the direction of travel by the meter pointer, but most will be read as I have described.

There are some precautions you should observe when using the multimeter: If you are not certain of the value of voltage or current you are going to measure, always place the selector switch on the highest value, then move it downward until the correct value is reached. If the meter is selected first on the low value and the actual value is higher, it may result in the pointer being broken or bent. Always recheck your meter settings before connecting to a live circuit. Be sure the polarity of the leads corresponds to the polarity of the circuit under test.

If the meter cannot be calibrated on either the high or low scales, this might be an indication that the batteries are weak and should be replaced.

When you have finished using the meter, make sure the switch is turned either to voltage or current. Never leave the meter on the resistance range, as this will decrease the life of the batteries.

Remember, when you are reading voltage, your meter should be in parallel with the circuit. When you are reading current, your meter is always in series with the circuit. Refer to Figure 1-2 (A) and (B) for example and explanation. Remember, when you are measuring resistance in a circuit, you must disconnect one end of the resistor, and always have the power turned off.

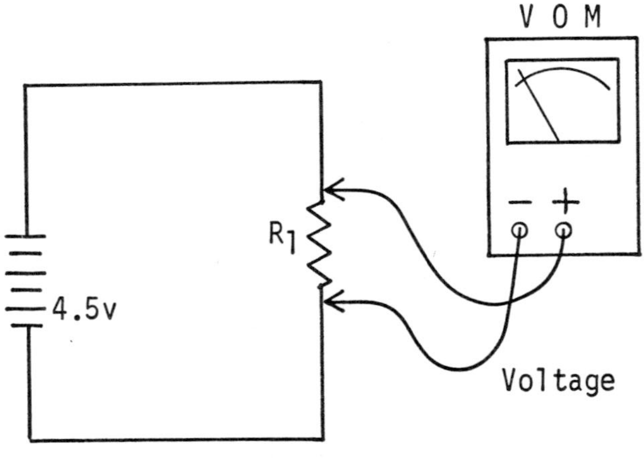

FIGURE 1-2(A): Illustration of how meter is connected across resistor to read voltage, the meter in parallel with the supply voltage.

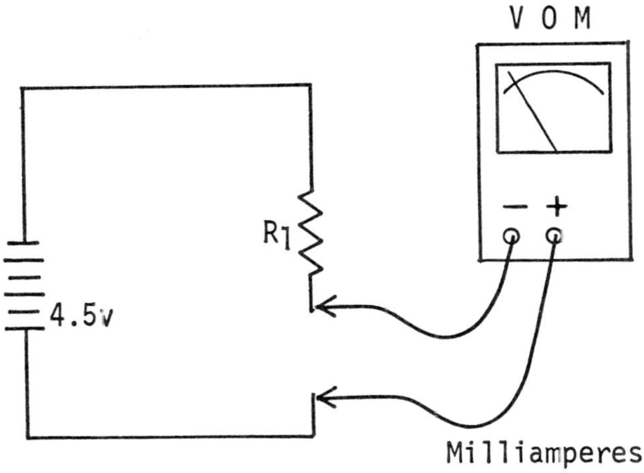

FIGURE 1-2(B): Illustration of how meter is connected to read current, the meter in series with the supply voltage and the resistor.

The Vacuum Tube Voltmeter

The VTVM is the way we usually refer to the vacuum tube voltmeter, which has this advantage over the VOM—the meter is isolated from the circuit by some form of vacuum tube bridge circuit. This means that the input impedance remains almost constant, from the low range reading to the highest range reading. With the VOM, the input resistance changes with each selection made on the selector switch. This is very important in some circuits (oscillator), because the meter resistance adds to the circuit, loading it down, which causes an error in the voltage readings with the VOM. Another very important advantage in the VTVM is a wider frequency coverage.

The multimeter AC voltage range increases in error at an approximate rate of .5 to 1% per 1000 Hz. However, the VTVM is fairly accurate above 10 MHz. The VTVM is almost a necessity in troubleshooting transistor and digital equipment. Another advantage is the polarity-reversing switch on the VTVM. In most modern equipment the chassis can be either positive or negative. It is not necessary to change the leads; simply switch the function switch to whichever polarity may be necessary.

Figure 1-3 illustrates a typical VTVM. The same precautions as mentioned in the use of the VOM must be taken when using the VTVM. Unlike the VOM, the VTVM will not read current, and since

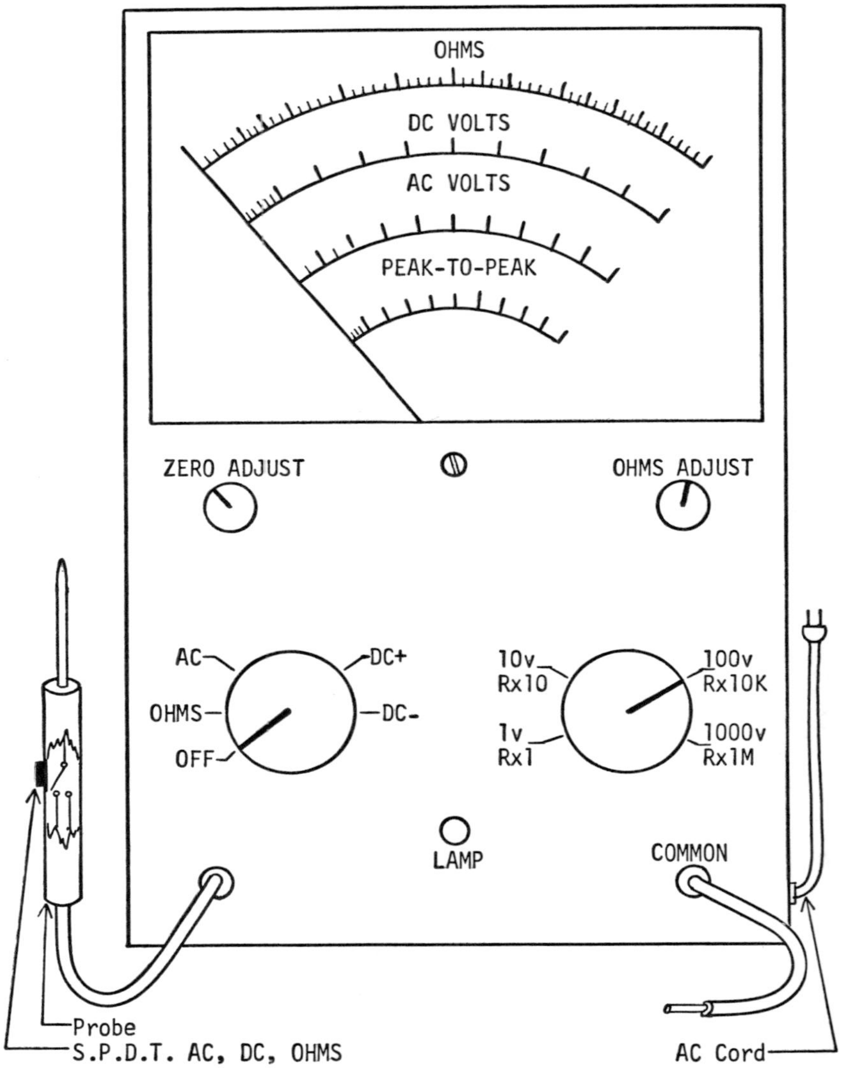

FIGURE 1-3: Typical vacuum tube volt-ohm-meter (VTVM).

it is plugged into the line voltage, it must be calibrated on all scales before using. When you first plug the meter into the line socket, let the meter warm up for a few minutes, adjust the zero control to zero, then connect the probes together and adjust the resistance control so the needle comes to the end of the scale. You may have to go back to the zero adjust after you have disconnected the probes and make a final adjustment. In some areas where the line voltage is changing, you may have to readjust the zero adjust knob several times while using the meter, if the line voltage varies too much.

The AC scale reads the RMS value of the sine wave, which is .707 times the peak value. Most of the modern VTVM meters will read peak-to-peak values fairly accurately. This is illustrated in Figure 1-3 as peak-to-peak (P-P).

The Audio Signal Generator

The audio signal generator is referred to much of the time as the AF generator. This instrument is an oscillator circuit, the frequency of which can be varied through the audio range from 20 Hz to 20 KHz, which is the audio frequency range. Most audio generators go beyond this range, down to 5 Hz and up to the ultrasonic frequencies or around 600,000 KHz. Not only can the frequency be varied, but the amplitude of the output signal can be varied. Some equipment using transistors

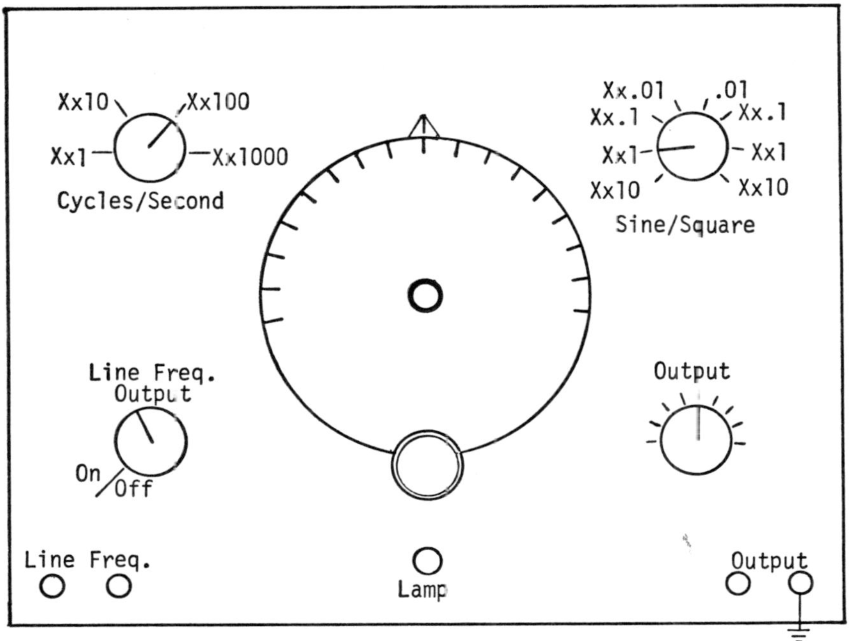

FIGURE 1-4: Audio signal generator.

and IC chips can be damaged by applying too large a value of signal voltage from the AF generator. When using the AF generator for signal injection, keep the signal level as low as possible to prevent damage to the equipment's components that are not defective to begin with.

The AF generator is used primarily as a signal injection into defective equipment operating in the AF range. The procedure usually starts either at the input or output, depending upon whether the equipment's entire circuit, or only part of it, operates in the audio range. The injected signal is traced either with an oscilloscope or a signal tracer until the defective stage is located.

Audio generators usually produce a square wave as well as a sine wave. In checking audio systems, sometimes distortion may be detected only by using a square wave.

Figure 1-4 illustrates a typical audio signal generator.

The Radio Frequency Generator

The radio frequency generator usually is referred to as just the RF generator. In principle this generator is about the same as the AF generator, the difference being that the RF generator frequency begins around 200 KHz on the low end of the range and extends to several hundred thousand MHz at the high end. Special types of RF generators may extend to several hundred thousand MHz.

Some RF generators have a steady audio signal that can be applied to the carrier for modulation. Other types have a pulse rather than a steady carrier for modulation. Figure 1-5 is a typical Rf generator. The selector switch shown at the bottom right hand corner will select a carrier with or without modulation.

The RF generator, like the AF generator, is used for signal injection through the RF stages. Start either at the input or the output RF stage, and use either an oscilloscope or a signal tracer to find the defective stage.

The RF generator is used also for aligning the RF and IF stages in an AM, FM, or TV receiver.

The Signal Tracer

Figure 1-6 illustrates a typical signal tracer. Basically the signal tracer is an audio amplifier, with provisions for both audio and RF input. When tracing through the audio stages, the probe is connected to

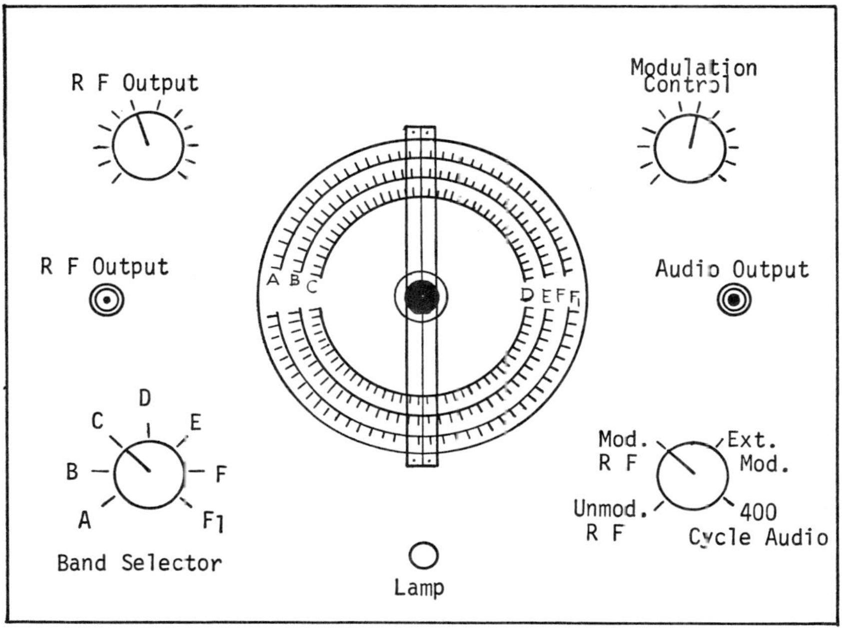

FIGURE 1-5: RF signal generator.

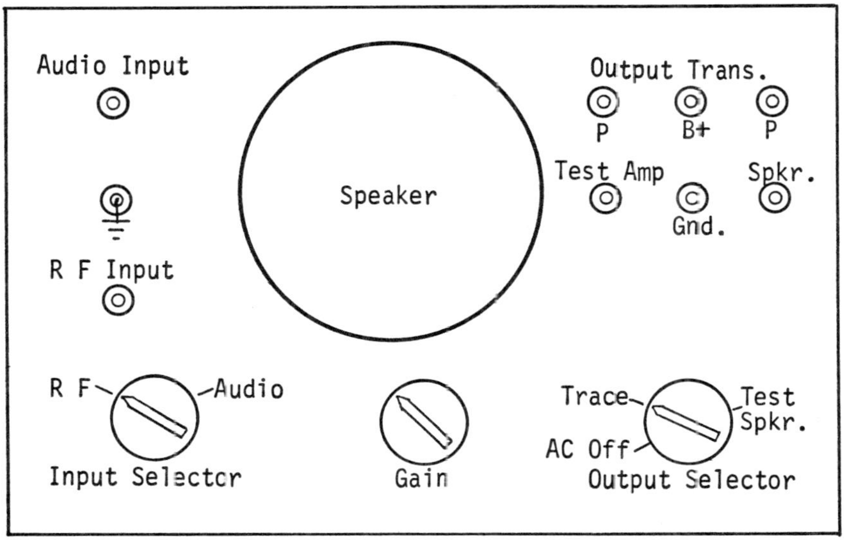

FIGURE 1-6: Typical signal tracer.

the audio stages one at a time until the defensive stage is found. Using the RF probe for the tracer, the same procedure is followed through the RF stages.

There are other convenient features of the signal tracer. A jack on the front panel disconnects the internal speaker, so that it may be used as a substitution speaker. Jacks are provided on the front panel for using the audio output transformer as a substitution transformer. Also a gain control is provided on the front panel to control the output of the AF and RF signals.

The Oscilloscope

Figure 1-7 illustrates a typical oscilloscope, referred to most of the time as just "scope." This instrument is used to look inside an electric circuit and display on the CRT (cathode-ray-tube) a picture, trace or wave form. This instrument is as useful to the electronic technician as the X-ray is to the physician. Both tell the operator just what is going on inside.

Oscilloscopes vary as to the number and placement of the controls on the front panel. But once you have learned to use one type and to interpret the wave form on the CRT, you should be capable of using any type of oscilloscope.

Troubleshooting modern electronic circuits requires a triggered dual-trace oscilloscope for best results. This type of oscilloscope will display two different wave forms at the same time. The advantage is that the input and output of a stage or complete circuit can be monitored at the same time, and a comparison made of the two.

Another consideration when choosing an oscilloscope is this: will it cover the band width you will be using the most? Oscilloscopes range from around 200 Hz to 14 GHz. If you are in the TV service business, it will not be necessary to buy an oscilloscope with a 30 MHz band width. All that is required is around 5 MHz for TV servicing. But if you are going to service micro-wave or radar equipment, then you should have an oscilloscope that will cover a wider band width. Servicing modern devices using pulse operated circuits and IC chips, you may find it to your advantage to choose a triggered oscilloscope and one with more than 5 MHz band width.

Before long, don't be surprised to see your modern oscilloscope in a case less than half the depth of the one you are now using. With the

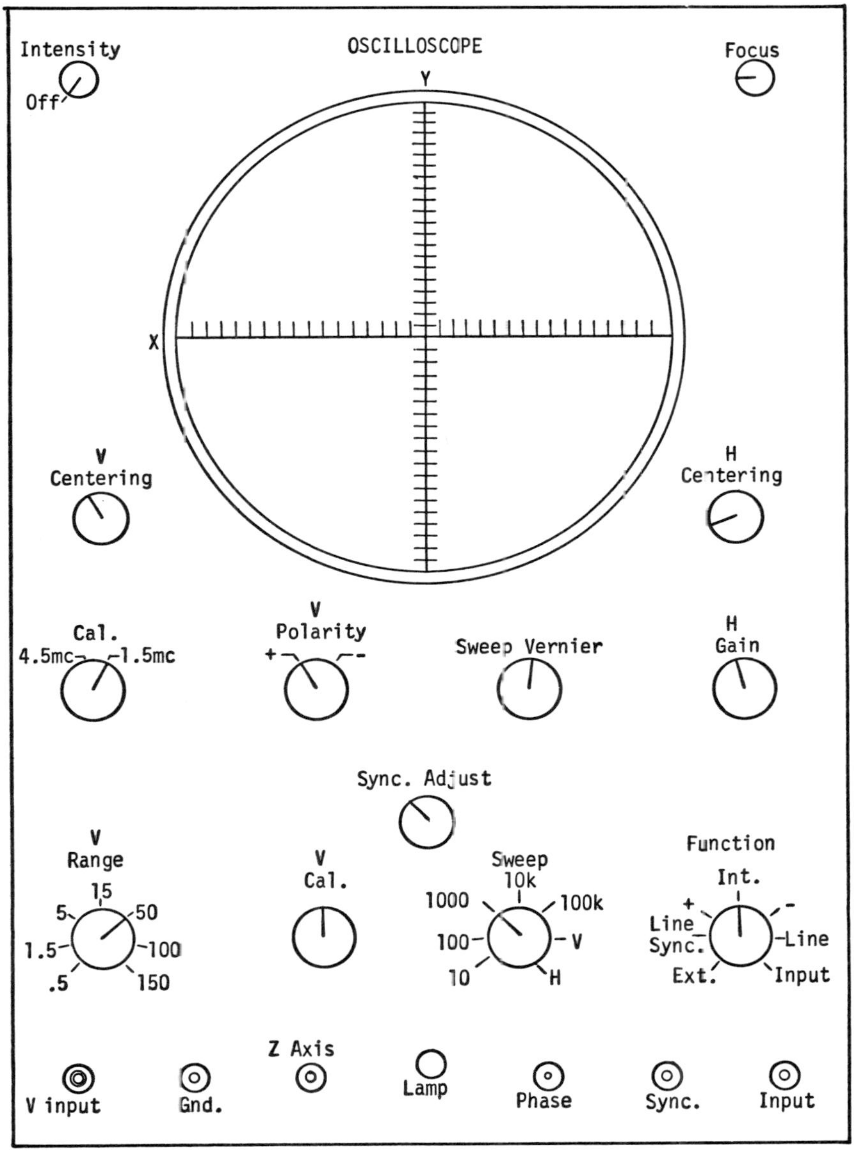

FIGURE 1-7: Typical oscilloscope.

rapid advancement that is being made in the LED (light emitting diodes), this could become possible.

Remember, be familiar with the face of your oscilloscope; it is telling you what is wrong inside the circuit.

The Resistance-Capacitance Substitution Box

This device may not be considered a piece of test equipment but it is essential when used in conjunction with your test equipment.

Figure 1-8 illustrates a typical resistance-capacitance substitution box. This device is made up of a variety of values of resistors placed on a rotary switch. The values are marked on the outside and a knob with a pointer indicates the values in sequence from the very low values to the very high values. A switch is usually placed on the front to select between the high and low values. A variety of capacitors are placed on a rotary switch also. The values are marked on the outside, and a pointer knob indicates the values.

The instrument is very useful when building new circuits. Instead of soldering and unsoldering components, connect the leads from this

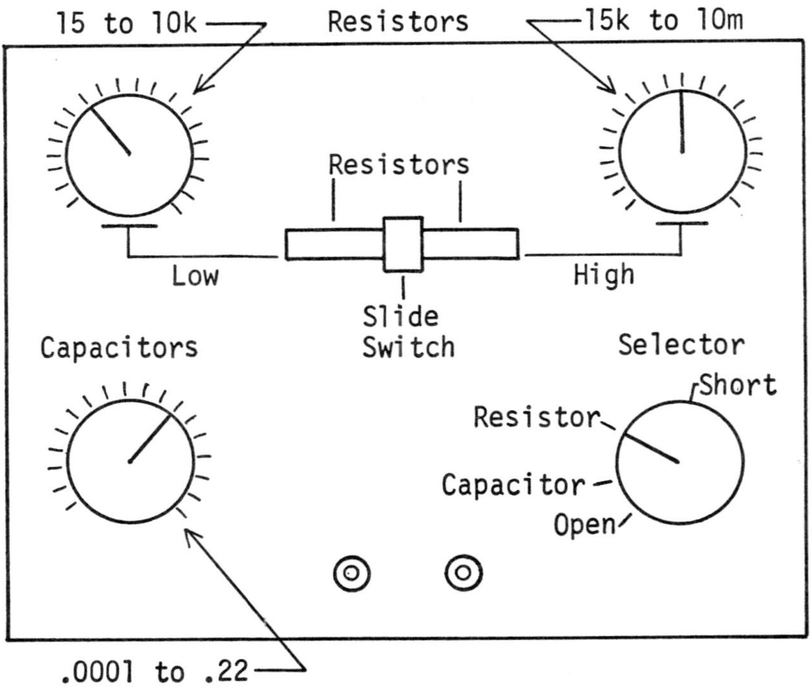

FIGURE 1-8: Resistor-capacitor substitution box.

instrument to your circuit and select the value that performs most effectively.

BEST WAYS TO OBTAIN TEST EQUIPMENT

When you decide to obtain a new or used piece of test equipment, there are a few things you should decide first. Let us consider, for instance, that you are in the TV service business.

- First, are you going to service all makes of receivers, or just one make alone?
- Second, are you going to service TV only or are you going to include other electronic devices?
- Third, will all of your service be performed in your shop or do you expect to perform service in the customer's home also?

These and other questions will determine what type of test equipment you will need. If you are going to service in the customer's home, you will not only need to equip your shop but you will need portable equipment also. If you are going to repair other electronic equipment, you will need a greater variety of test equipment. If you are going to repair only one make of TV, you will only need equipment necessary for repairs for that one make.

Remember, this equipment is going to work for you, so choose the equipment that will be the most effective and that will not be obsolete in a few years. For instance, in choosing a VOM or VTVM, why not consider the digital? It is true it may cost a little more, but it will not be obsolete in a few years and it can be used on the older devices as well as on the latest. It is easier to read because of the larger displayed numbers, and its accuracy is greater than that of the ordinary meter. Look over some of the advantages offered by this and some of the other digital equipment. Look forward to the future and try to make sure that the testing equipment you choose will be adequate in its use.

If you are in industry and have to choose the test equipment that you or someone else will use, first decide what this equipment will be called upon to perform. Will it take care of today's needs, and if any changes are made, will it perform under these new changes?

If you are a hobbiest and choosing test equipment, consider that your instrument may be called upon to perform all types of tasks. Do not purchase a meter that only reads 1000 volts maximum and then a few days later find that you have to measure 5000 volts. Do not purchase a general purpose oscilloscope, and then, when you are designing a new circuit, find that your oscilloscope will not measure DC. If you are an amateur radio operator, do not purchase a digital frequency meter without some provisions for measuring into the UHF. Today your highest frequency of operation is 30 MHz, but next week you hear how much fun it would be to be operating 144 MHz through a repeater, and you are stuck with a frequency meter that will only measure up to 30 MHz.

HOW TO DETERMINE WHETHER TO PURCHASE FACTORY ASSEMBLED EQUIPMENT OR KITS

Your choice of factory assembled equipment or kits that you assemble will depend largely on who you are buying them for and where this equipment is to be used. If you are buying for industrial use, you will buy the best possible equipment that will withstand very rugged use. You are going to buy factory assembled equipment because this type of equipment does not come in kit form.

If you are a hobbiest, then you might consider some of the advantages and disadvantages of kits. For instance, there are some very good equipment kits on the market. In most cases they are accurate enough for most experiments. They are sturdy, and if you know something about using the soldering iron and can follow instructions, you can do a good job in assembling the kit. You will gain some knowledge in using parts and assembling them into a circuit. Some kits, when you assemble them yourself, may save you up to $150.00.

If you are going into the service business, you may not want to take the time to construct a kit, and perhaps you are not sure you can assemble a large kit. If you made a mistake and couldn't find the trouble, you would have the bother and the expense of sending it back to the factory. However, most kits come with a very complete instruction book, and if you are careful to follow the instructions you are not likely to make a mistake. Remember, most of the trouble found in the assembly of kits is poor solder joints, so if it is your first kit, take your time and be very careful in soldering.

In choosing between a factory assemble and a kit, there are many advantages and disadvantages. Your choice should be according to your needs, what it will be used for, and of course the money available to purchase it.

2

How the Vacuum Tube Operates

VACUUM TUBES CLASSIFIED ACCORDING TO NUMBER OF ELEMENTS

In the field of electronics the vacuum tube is an important device because of its ability to detect, amplify, and rectify an electric voltage or signal. Its ability to accomplish this is due to the placement and number of elements.

During Thomas Edison's research with the incandescent lamp he discovered that if a metal plate was placed near the filament inside the bulb, current would flow between the plate and the filament if the plate was made positive with respect to the filament. If the plate was made negative, no current would flow. This was the diode rectifier, or two-element tube, that would perform a certain function—i.e., rectify.

Early in 1900 Lee De Forest discovered that if a grid of wire was placed between the filament and the plate inside the bulb, current could be controlled, if the potential on the grid was varied.

From a very simple incandescent light bulb, and by placing the elements in the proper position, the modern vacuum tube has come into being. With the proper polarity attached from the outside, it will perform many functions.

Modern vacuum tubes are classified according to the number of elements comprising them. The following list shows the tube by number of elements, and the name of the elements:

Tube Name	*Number of Elements*	*Elements Name*
Diode	Two	Cathode and plate
Triode	Three	Cathode, grid, and plate

Tetrode	Four	Cathode, grid, screen grid, and plate
Pentode	Five	Cathode, grid, screen grid, supressor grid, and plate

An explanation of the purpose and use of the different elements will be given later in this chapter.

A REVIEW OF CURRENT FLOW

Before proceeding further into vacuum tube theory, let us briefly review the theory of current flow. You have learned earlier in your study of electronics the definition of current flow:

> *Current:* The rate of transfer of electricity from one point to another. Current is usually a movement of electrons but may also be a movement of positive ions, negative ions, or holes.

An electron is a negatively charged particle surrounding the nucleus of an atom. Remember, like charges repel and unlike charges attract. The electrons are constantly moving in an orbit within the atom, but those electrons in the outer orbit or valence are the only ones free to move away from the atom. The force holding the electrons in their orbit varies with different types of atoms. In some conductors electrons are free to move about within the conductor. (These are called "free electrons.")

An atom from which an electron (negative charge) has moved away may be considered to have a positive charge.

If in a solid conductor an electron moves free of the atom, it may move into the positive ion of the next atom, leaving the previous atom with a positive charge. An electron leaving another atom may move into this positive ion, etc. The moving of electrons from one atom to another makes it appear that electrons are moving in one direction, and the positive ions or *holes* are moving in the opposite direction.

The flow of current in a solid conductor is a flow of electrons. If the conductor is heated by applying a sufficient amount of voltage, the electron will gain enough energy to leave the material. This is called *thermionic emission.*

> *Thermionic emission:* The liberation of electrons from a solid as a result of heat.

CURRENT FLOW IN THE VACUUM TUBE

The emitter of electrons in the vacuum tube is called the cathode.

There are two types of cathodes used in the vacuum tube, the *directly heated* and the *indirectly heated*. Figure 2-1 shows the symbol

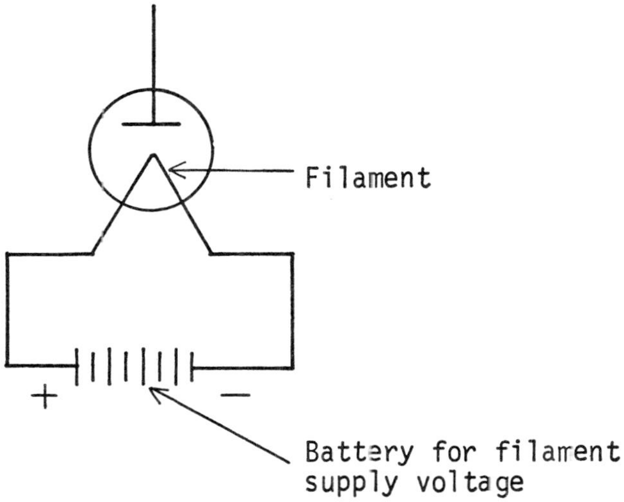

FIGURE 2-1(A): Filament or directly heated cathode vacuum tube; battery is used to heat the filament or cathode.

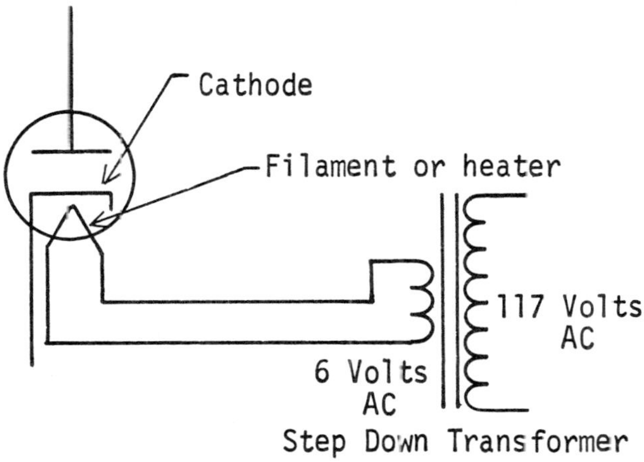

FIGURE 2-1(B): Indirectly heated cathode; the filaments are connected to a step-down transformer which is connected to the AC line. In the top of the circle is the symbol for the plate.

used for both types: (A) shows the filament or directly heated type; (B) shows the heater (cathode) or indirectly heated type. In (A) the filament or directly heated cathode is connected to a battery which supplies the necessary voltage to heat the filament; the second symbol in the top of the circle is the plate or anode. In (B) the heater or indirectly heated cathode is connected to a transformer which is connected to an AC source. This is a stepdown transformer to supply the necessary voltage for the filaments.

The heater is a spiral-shaped conductor mounted inside a metal cylinder called the cathode. The cap-shaped line placed over the heater is the symbol for the cathode, and the symbol at the top is the plate or anode. An AC voltage source is usually used to heat the filament or cathode because it is generally available and less costly than DC from batteries which have to be replaced frequently.

The filament or heater in the vacuum tube is not considered one of the elements of the tube. Its only purpose is to heat the cathode, to cause electron emission. Since the indirectly heated cathode is more widely used in modern electronics, we will use this symbol for our study of the vacuum tube; and since the heater is not considered one of the elements, we will omit it in our diagrams. We will refer to the cathode as the emitter of electrons.

Figure 2-2 shows the symbol of the diode vacuum tube or the two-element tube. These elements are cathode and anode (or plate). The diagram shows the cathode with a negative sign beside it, and the

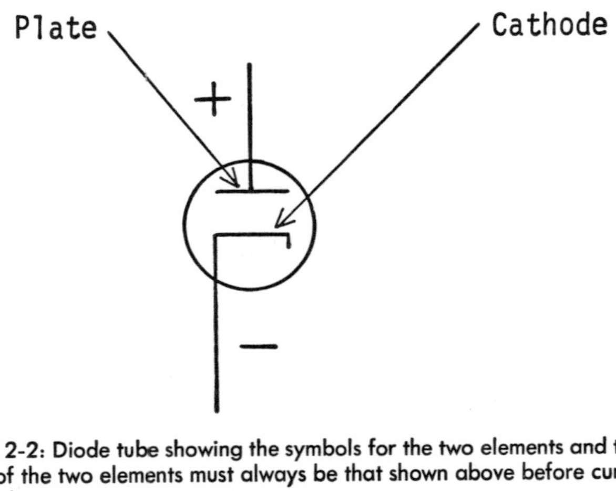

FIGURE 2-2: Diode tube showing the symbols for the two elements and their polarity. The polarity of the two elements must always be that shown above before current will flow through the tube.

plate has a positive sign beside it. Before current can flow in the vacuum tube, the cathode must be at a negative potential and the plate at a positive potential. If the plate were to become negative no current would flow through the tube. I will show you how to prove this statement.

Construct the circuit in Figure 2-3 (A). You will need an eight-pin tube socket, a 5u4 tube (it has two plates, but we will only use one for this experiment), a six-volt battery to heat the filaments, a six-volt battery for the power source, and a VOM.

Connect one battery to the socket pins 2 and 8, which are the filaments. Connect the other battery and meter as shown to the proper terminals of the socket. The meter should be switched to read current on the highest scale first. Be sure the polarity of the meter is the same as that shown in the diagram. With this circuit you will prove that if the plate is made negative and the cathode is positive, there will be no current flow in the circuit. Second, you will prove if the plate is positive and the cathode is negative, current will flow through the circuit.

If you have the circuit connected as shown, you may notice the meter is not moving. Switch the meter to a lower range and notice the

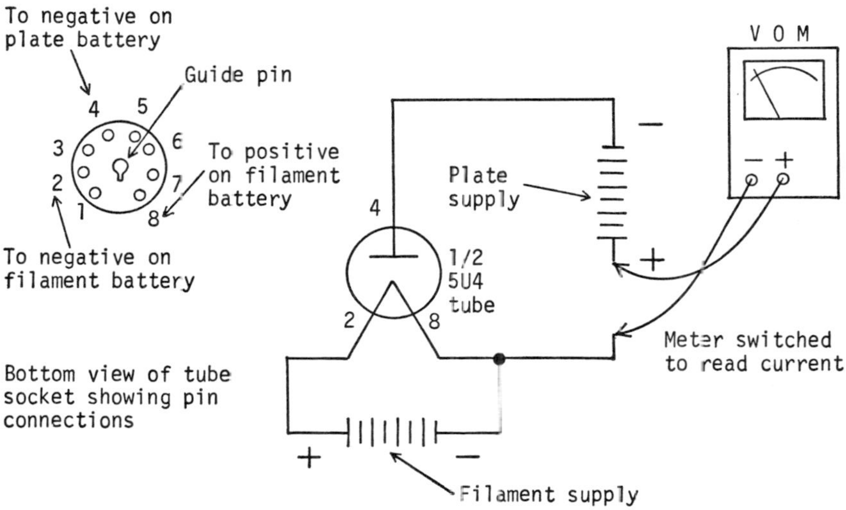

FIGURE 2-3(A): Diode tube with polarity of elements reversed; plate is at a negative potentilal, and the cathode is at a positive potential. Notice meter will not read current.

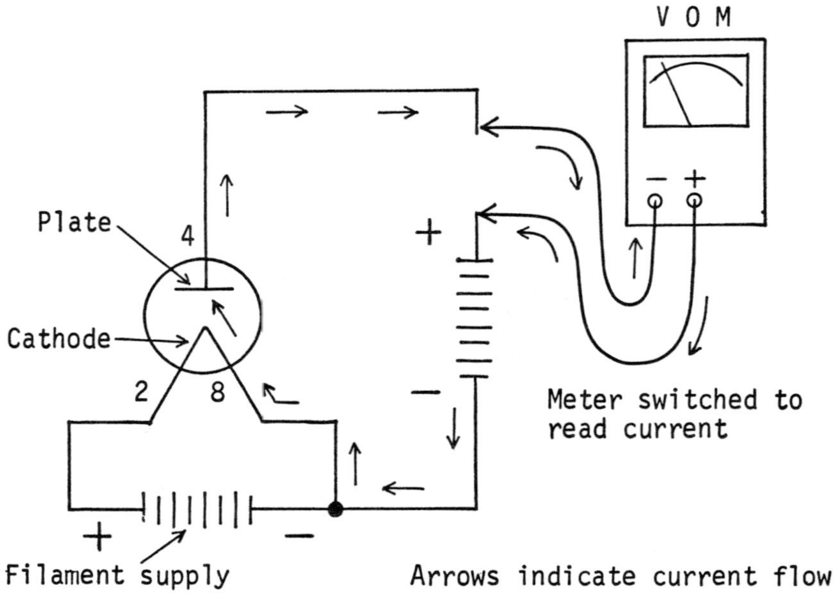

FIGURE 2-3(B): Diode tube with supply voltage connected correctly. The plate is at a negative potential. Notice the meter is showing some current flow. The current path is indicated by the arrows, current flowing from the cathode through the tube to the plate.

meter is not reading. Continue to lower the range switch to the lowest range. Notice the meter is not reading even on the lowest range. This proves that no current is flowing when the plate is negative and the cathode is positive.

Refer to Figure 2-3 (B) and connect the battery as shown; do not connect the meter until you have reversed the leads for proper polarity. The circuit now is reversed: the plate is positive, and the cathode is negative. The meter should be switched to the highest range. If the meter is not reading, switch down to the next range or ranges until the meter is reading current. With the meter reading current you have proven that if the plate is positive and the cathode is negative, current will flow through the tube and circuit.

Refer now to Figure 2-3 (C) and install a 600 ohm resistor in series with the plate and the supply. Connect the meter as shown, making certain to observe the correct polarity before applying power to the circuit, and switch the meter to the highest current scale. Apply power; if the meter does not read, switch to a lower scale until it does read. Notice the meter is reading a higher value of current than that in Figure 2-3(B). The resistor represents a load that would be connected

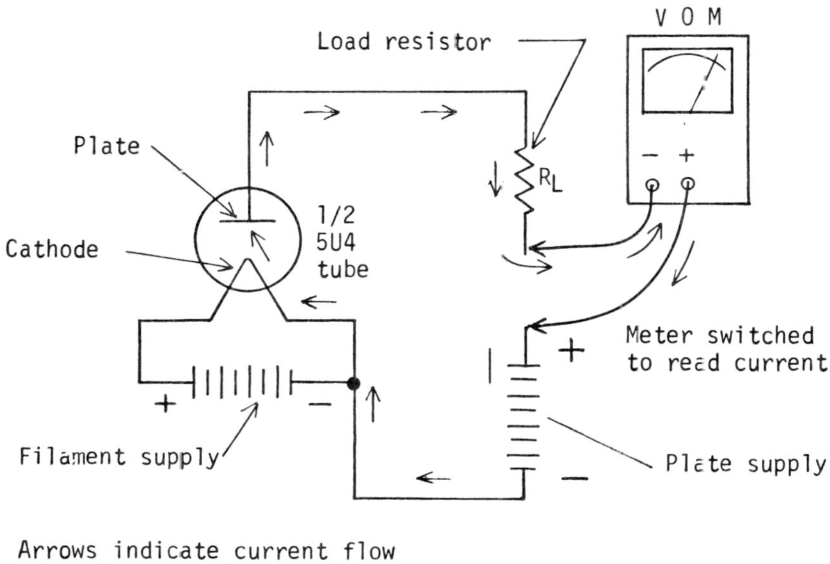

FIGURE 2-3(C): Diode tube with load resistor connected in series with plate. Notice that with the plate connected to a load more current will pass through the tube, as indicated by the increase in meter reading over that shown in (B), which has no load other than the meter resistance.

to the tube while it is in operation. Try different values of resistors in place of the 600 ohm, and notice as the load is increased the current reading on the meter will also increase as long as the supply voltage is held the same.

Diode vacuum tubes are used in two applications—rectification and detection. These two applications will be discussed later.

CONTROLLING CURRENT FLOW IN THE VACUUM TUBE

Current flow through the vacuum tube is from the cathode to the plate. How can we control this current through the tube? In the diode, or two-element tube, if you increased or decreased the voltage applied to the plate the current also increased or decreased. Current through the tube was controlled by the voltage. Increasing the plate voltage increases the plate current; all of the electrons emitted from the cathode are drawn to the plate. Increasing the voltage still farther, a point will be reached when the plate cannot hold all of the electrons emitted from the cathode. Plate current will cease; a point is reached called *saturation*. Plate current for this condition is referred to as *saturation current* or plate saturation.

The Triode Vacuum Tube

In the first part of this chapter, I mentioned it was Lee De Forest who first found that by placing a wire grid between the cathode, and the plate inside the tube, current passing through the tube could be controlled if the potential placed on the grid was varied. This construction inside the tube is used in modern tubes for controlling the current flow through the tube.

When voltage is applied to the tube a stream of electrons flows from cathode to the plate. A grid is placed directly in this electron path. In structure, the grid is a fine mesh of wire, and if enlarged it would look similar to the screen found in screen doors and windows. It is positioned nearer the cathode than the plate. The reason for this is it will have more control if it is closer to the source. Most of the electrons leaving the cathode flow through these tiny holes and onto the plate. Not all of the electrons emitted from the cathode reach the plate; some lodge on the grid structure. If the grid were not connected to some potential it would be left "floating," electrically. As these electrons that lodge on the uncharged grid begin to build up, the grid becomes more and more negative. As this negative charge begins to build up it starts repelling electrons emitted from the cathode. If the grid were left "floating," it would not control the current through the tube but would cut it off. The grid placed in the electron stream is called the *control grid*. In order for this grid to control the current, it must be "tied"

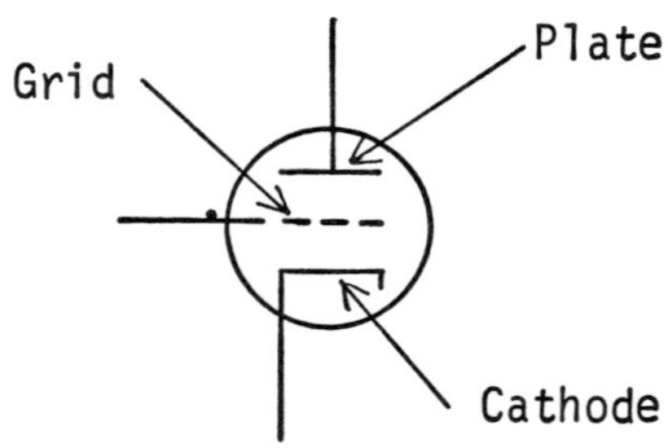

FIGURE 2-4(A): Triode vacuum tube showing element placement and their names.

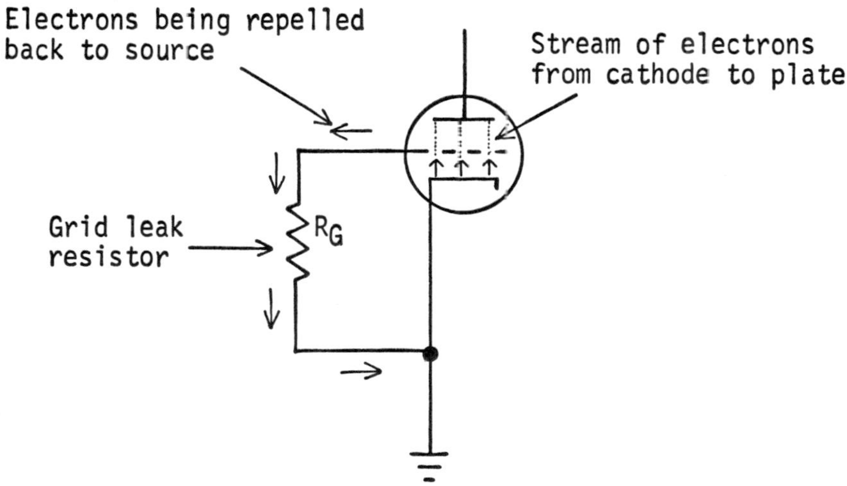

FIGURE 2-4(B): Triode tube showing the grid leak resistor connected between the grid and ground. Notice the stream of electrons flowing from cathode to plate through the mesh grid holes. The arrows coming out of the grid down to ground are those electrons that collected on the grid and are being repelled off it down the path formed by the resistor.

back to the cathode to return these collected electrons. In practice a grid-load resistor is connected from the grid to ground, which forms a path back to the cathode.

Figure 2-4(A) shows the symbol for the triode tube with the elements marked, and Figure 2-4(B) explains the current flow through the tube and the grid resistor connected between the grid and ground. Notice, most of the current reaches the plate, while some of the electrons are forced along the path to ground through the resistor.

Grid Bias

This term is defined as a DC voltage applied to the control grid electrode to establish the desired operating point.

There are two states or conditions in which a vacuum tube is placed. The first is the *static state*. In this condition, the necessary DC voltage and bias (control grid voltage) and plate load (resistor or inductor) have been connected. The tube is drawing current, but it is a steady current controlled by the grid bias. The tube is just "idling," not producing an output.

The second condition is the *dynamic state*. In this state, an AC signal voltage varies from its positive alternation to its negative alternation; the operating grid bias is changing. This change in the grid voltage causes a change in the current flow between the cathode and the plate. This will be discussed later in this chapter.

Figure 2-5(A) shows a vacuum tube with a resistor connected as a load in the plate circuit, and a resistor in the grid circuit as a load. Figure 2-5(B) shows an inductor connected as a load in the plate circuit, and a grid load resistor. These two circuits represent different types of loads that may be used in the plate circuit.

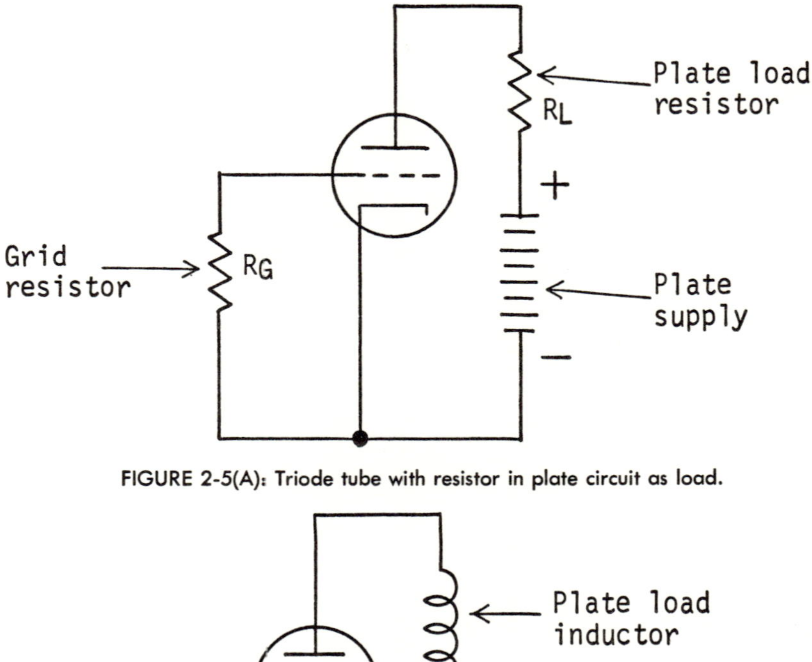

FIGURE 2-5(A): Triode tube with resistor in plate circuit as load.

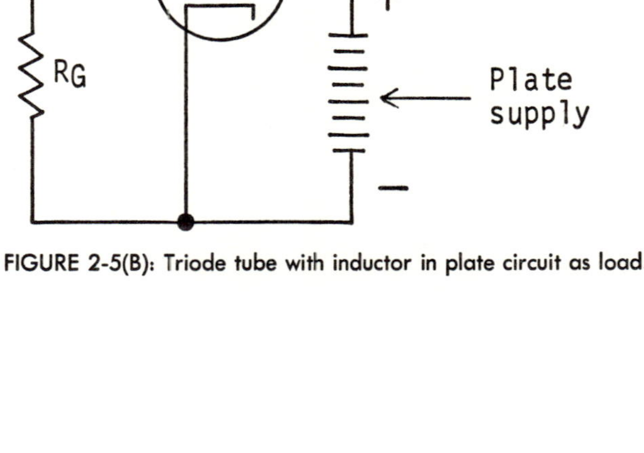

FIGURE 2-5(B): Triode tube with inductor in plate circuit as load.

There are three types of grid bias: grid-leak, fixed, and cathode (sometimes called protective bias). Figures 2-6(A), (B), and (C) illustrate the three types.

Figure 2-6(A) illustrates grid-leak bias, also referred to as automatic bias. In a practical circuit, the coupling capacitor would be connected to the grid (not shown). During the first 90° of the positive alternation of the AC signal applied through the capacitor to the grid, the grid goes positive. The positive grid will cause the plate current to increase as the grid draws current. This grid current will charge the capacitor *negative* on the side connected to the grid, and *positive* on the opposite end. During the second 90° of the positive input signal alternation, the signal will be less positive and the grid voltage will drop below zero. The capacitor will start discharging from the negative side down through the grid resistor, whose value is usually very large. This will place a negative voltage on the grid, and plate current will decrease. During the negative alternation of the input signal the grid will remain at a negative potential. However, during the fourth 90° of the signal input, the voltage is going more positive and the plate current will increase back to its starting value.

Figure 2-6(B) illustrates fixed bias, which can be a battery, as shown, or a DC power supply, that can be adjusted for the proper value

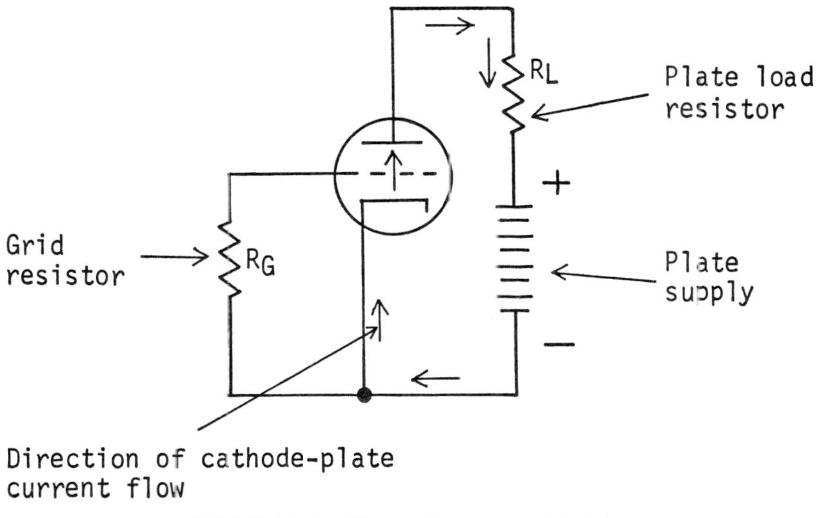

FIGURE 2-6(A): Triode tube using grid leak bias.

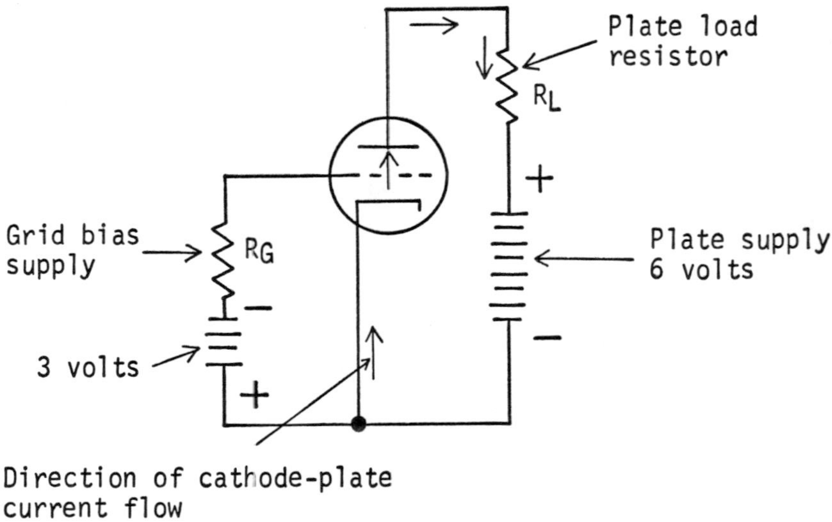

FIGURE 2-6(B): Triode tube using fixed bias. Notice that the negative side of the battery is connected to the ground. Notice also that the value of the grid voltage is less than the plate voltage; this is normal for tube operation.

of grid bias. Notice that the grid is connected to the negative side of the battery. The grid is always negative with respect to the positive cathode, or the cathode is positive with respect to the negative grid.

Figure 2-6(C) illustrates cathode bias. The bias on the grid of a vacuum tube is always measured between the control grid and the cathode, the grid being negative and the cathode positive. If a resistor is placed in the cathode circuit, the current passing through the tube must pass through this resistor and cause a voltage to be dropped across it. The voltage at the top end of the resistor will be positive: this voltage is the bias for the tube. The capacitor C_k is called the cathode bypass capacitor and serves two purposes. First, it bypasses any AC signal in the cathode circuit to ground, or bypasses the AC around the power supply for the grid (voltage drop across the resistor). Second, it acts as a voltage regulator for the grid bias supply, to keep it constant (capacitor's ability to oppose any change in voltage across it).

A way to gain a more effective understanding of how plate current is controlled by a changing grid bias would be for you to observe it in an operating circuit. Construct the following circuit using a 6J5 triode tube, a plate load resistor, and a cathode bias resistor. You should have a power supply capable of supplying approximately 300 volts DC.

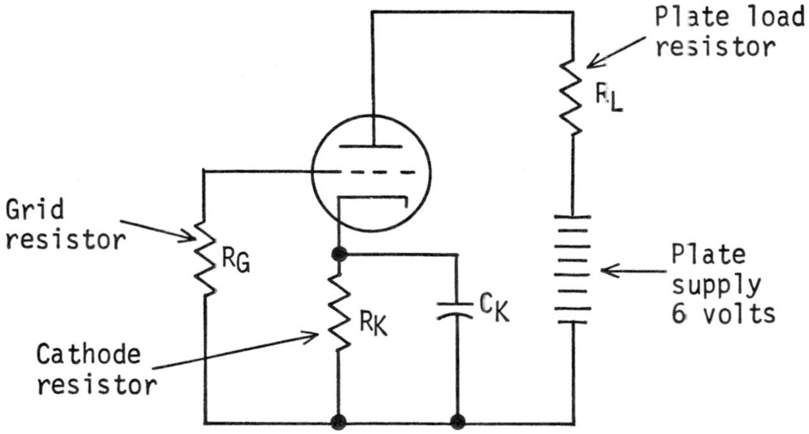

FIGURE 2-6(C): Triode tube using cathode bias. The component CK is the cathode capacitor; its function will be explained later.

How do you find the correct value of voltage, current, and resistance to use with this tube?

First consult your tube characteristic manual for plate voltage, plate current, and operating grid bias. You should also look at the base diagram for this tube to find what elements of the tube are connected to what pins on the base of the tube socket.

In the manual you find the following information: E_p plate voltage is 250 volts, I_p plate current is 9ma., and E_g grid voltage is -8 volts.

Now assume that you have a supply voltage (E_{bb}) of 300 volts. Find the value of the plate load resistor $-R_L$. The supply voltage is 300 volts, and the plate needs 250 volts of this 300 volts to operate. The resistor will have to drop the difference, or $300 - 250 = 50$ volts. The plate current is 9ma.; this current will flow through this resistor to drop the 50 volts. You know two values now, so find the third by substituting these values in Ohm's Law formula for resistance.

$R_L = \frac{E_{RL}}{I_p} = \frac{50}{.009} = 5555$ ohms; use the nearest standard value which is 5700 ohms. Now find the power this resistor must dissipate: $P = E_{RL} \times I_p = 50 \times .009 = .450$ watts.

In selecting wattage for resistors it is good practice to use a higher wattage value, so use a one-watt here.

Find the value of the cathode bias resistor. The voltage developed across this resistor is the grid bias, and the manual gives this value as -8 volts. The same current that passed through the plate load resistor must pass through this resistor also, so substitute these values in the formula.

$\frac{E_g}{I_k} = E_g = \frac{8}{.009} = 888$ ohms; use the nearest standard value which is 900 ohms. Now compute the power dissipated by this resistor: $P = E_g \times I_k = 8 \times .009 = .072$ watts; use a one watt.

Now that you have the values of the components, construct the circuit in Figure 2-7(A). The bottom view of the 6J5 tube is shown with the pin numbers and the name of the elements. This diagram shows a 6J5 triode tube using cathode bias, and meter readings shown are for the static condition of the tube. Remember, this is the "idling" condition, before signal is applied.

M_1, shown by the dotted lines, is a voltmeter, connected across the plate load resistor. It reads the 50 volts that you calculated for the voltage drop across this resistor. Observe the polarity of the meter

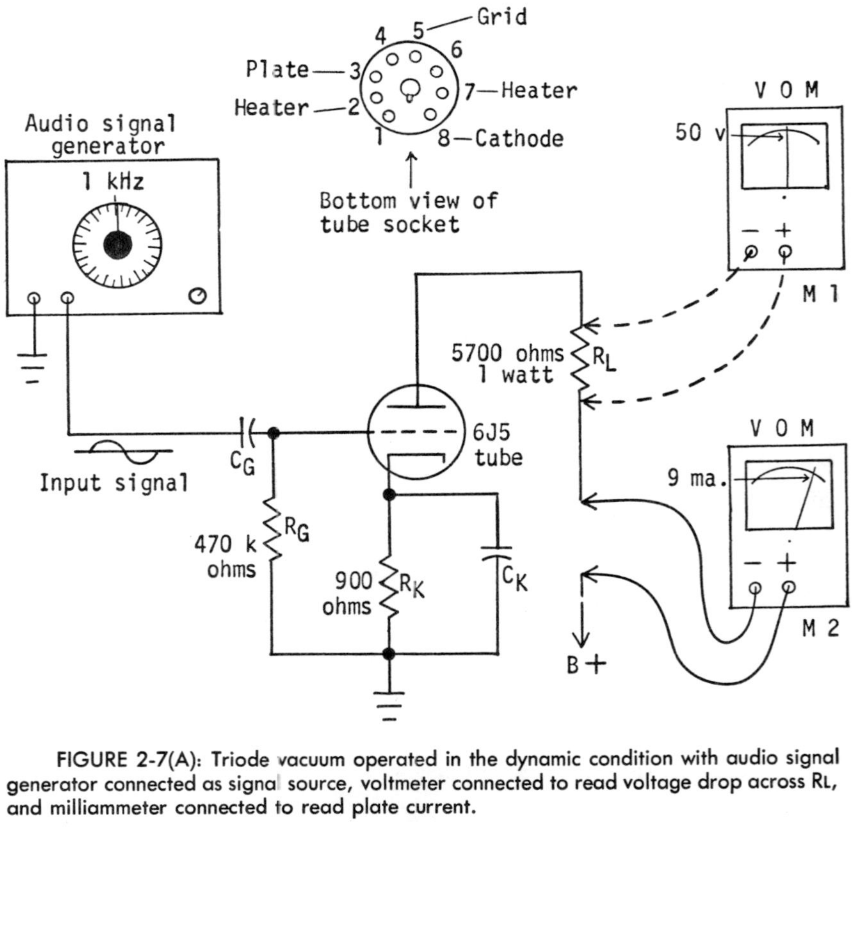

FIGURE 2-7(A): Triode vacuum operated in the dynamic condition with audio signal generator connected as signal source, voltmeter connected to read voltage drop across R_L, and milliammeter connected to read plate current.

leads. The top of the resistor is negative, and the bottom is positive; current flow is down through this resistor. Always make certain of the direction current flows before installing a meter. M_2 is a milliammeter connected in series with the plate load, to read the plate current passing through this resistor. Notice it is reading 9ma. This would be called the static plate current. Observe the polarity of the meter which is connected in series with the current flow.

The audio signal generator is coupled through a capacitor to the grid of the tube. This will simulate the dynamic conditions of the tube, or how a changing grid voltage will cause a change in the plate current.

Turn the audio generator on and turn up the gain control; notice, as you turn the control up, both meters will change values. The voltmeter across the resistor will start to increase, and the meter reading current will also start increasing. What is being simulated here is that during the positive alternation of the input sine wave, the grid is going more positive, or less negative. This causes the plate current to increase, as indicated on the meter. An increase in current through the resistor will cause a greater drop across the resistor, as indicated by the meter. If the drop across the load resistor increases this will cause the plate potential to decrease, so the plate is swinging less positive. If you were to connect M_1, with positive on the plate and the negative to ground, you would notice that the voltage is reading less than the 250 volts we started with. This then shows that the plate potential is swinging negative.

Turn the gain control on the generator down; this is to simulate the negative alternation of the input wave form. The grid is going in a negative direction, or swinging less positive. Notice the change in the readings of both meters: the voltage is decreasing, likewise the current meter is decreasing. This decrease in voltage and current at the load resistor will cause the plate potential to swing up, or go more positive.

While observing the readings on the two meters, did you notice, as the grid was swinging in a positive direction, that the plate was swinging in a negative direction? And that as the grid was swinging in a negative direction the plate was swinging in a positive direction? The signal on the grid and the signal on the plate are 180° out of phase with one another. This is true with all amplifiers, where the signal is fed to the grid and removed from the plate.

Refer to Figure 2-7(B), which is a portion of the circuit of 2-7(A). This illustrates how to connect your voltmeter to read grid voltage,

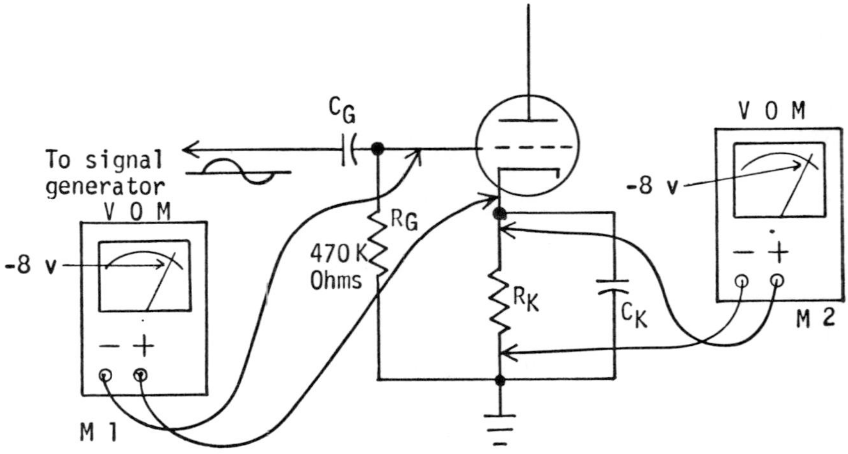

FIGURE 2-7(B): Illustration of how grid voltage can be measured. Grid voltage should always be measured between the grid and the cathode. Notice the polarity of both meters; the grid is always negative with respect to a positive cathode. Both meters should read the same.

and, where cathode bias is used, how the meter should be connected to read voltage drop across the cathode resistor. The readings shown are for static condition of bias voltage. The negative lead of meter M_1, is connected to the grid, while the positive lead is connected to the cathode. Recall that the grid is always negative with respect to the cathode. M_2 is connected with the positive lead at the cathode terminal, and the negative lead connected to ground. Both M_1 and M_2 should read the same voltage. This proves that the voltage drop across the cathode resistor is the grid bias voltage.

What you have learned so far in this chapter will give you a great advantage in troubleshooting electronic equipment using vacuum tubes.

First, you have learned how to calculate the value of some of the components used in an amplifier circuit.

Second, you have learned the proper method for measuring the voltage drop across a resistor, and the proper method for measuring the current in a circuit. You have observed how the voltage on the grid controls the current through the vacuum tube.

Third, you have observed that the signal voltage on the grid is 180° out of phase with the output on the plate.

Suppose you were working on a piece of electronic equipment

that used a 6J5 tube. You did not have, or could not obtain, a diagram for this equipment. You observe after inspection that the plate load resistor on this tube was burned so that you could not recognize the colors to replace it. By applying what you have learned, you could replace this part without difficulty.

Suppose you were working on an audio amplifier that used a 6J5 tube in one of the stages. The gain through this stage is low; you checked the tube and found it good. By applying some of the things you have learned in this chapter, such as measuring the voltage on the plate of the tube, measuring the voltage drop across the load resistor, measuring the grid bias, you could possibly repair this equipment even without a diagram.

Tetrode Vacuum Tubes

So far in this chapter we have dealt only with current flow through the tube. The elements in the triode tube are small pieces of metal which form a capacitance between these elements. This capacitance is formed between the cathode and control grid, and between control grid and plate.

When the triode is operated as an AF amplifier it is limited to around 30 MHz, due to the large interelectrode capacitance and transit time, which refers to the velocity or speed of the electron from the time it leaves the cathode until it reaches the plate. When the electron from the cathode reaches the control grid it is repelled; the grid is negative, and like charges repel, so the electron is slowed down. To reduce this interelectrode capacity and speed up the transit time, the second grid was installed. The construction of this second grid, called the screen grid, is very similar to that of the control grid. It is a fine mesh wire, whose small openings allow the electrons to pass through to the plate, but a positive potential is placed on this grid, and some of the electrons collect on it. When the control grid slows the electrons down by repelling them they pass on toward the plate. The positive potential on the screen grid attracts them, and they are accelerated on toward the plate. Some of these electrons are attracted on the screen grid, therefore current flows in the screen grid circuit.

Figure 2-8 illustrates a tetrode tube with the necessary components to make this tube operate. Two new components have been added. The first is R_{sg}, or screen grid dropping resistor, which is

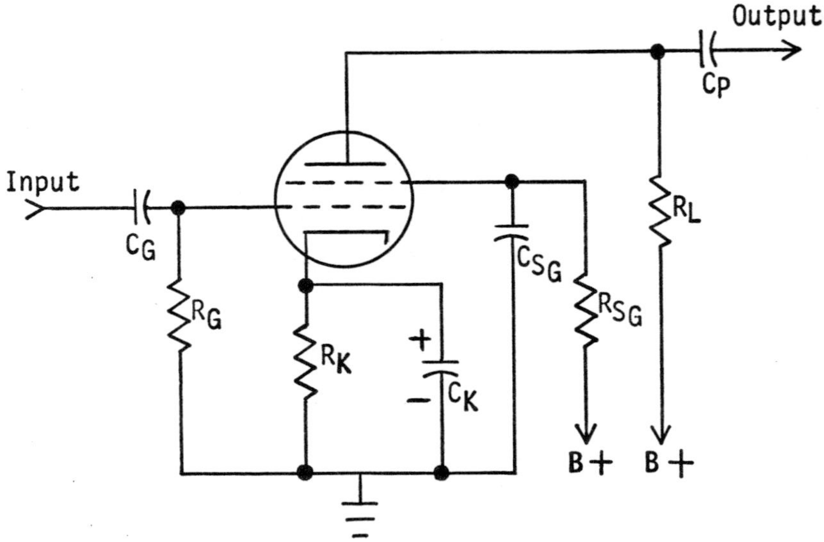

FIGURE 2-8: Tetrode tube amplifier circuit showing the symbols for the different components. The references for these symbols are as follows: C_G—coupling capacitor; R_G—grid resistor; C_K—cathode capacitor; R_K—cathode resistor; C_{SG}—screen grid bypass capacitor; R_{SG}—screen grid resistor; R_L—plate load resistor; and C_P—the plate coupling capacitor.

necessary to reduce the supply voltage to acquire the screen grid operating voltage. (This will be discussed later in this chapter.) The other new component is C_{sg}, which is the screen grid decoupling capacitor. When the tube is amplifying, the screen grid is directly in the path of the signal; therefore, it is necessary to couple this AC signal voltage around the screen power supply, which is the screen dropping resistor. The advantage this tube has over the triode tube is that interelectrode capacity has been reduced and transit time has been speeded up, both of which factors contribute to the ability of this tube to operate at much higher frequencies.

Pentode Tube

All of the electrons that reach the plate are not held there; some strike the plate surface with such force they cannot be held. These electrons bounce off and are free to roam in the space around the tube. Other electrons may not strike the plate surface at all; they also are free to roam around in the tube. Soon they begin to accumulate (a process referred to as space charge), and if allowed to build up would eventually start repelling the electrons coming from the cathode. To overcome the effects of space charge another grid is installed between the plate and screen grid. This grid is called the supressor grid; it is connected to the cathode either internally or externally, which causes it to be at a negative potential. The electrons that either did not strike the plate and hold there or the ones that missed the plate are repelled back to the plate by the supressor grid, because of its negative potential. Some do stick on this grid and find their way back to the cathode, since the supressor grid and the cathode are tied together. The electrons that were normally lost in the space charge are now added to the plate current, which increases because of this. The pentode tube then normally operates with a greater plate current and a higher gain in the output.

From the following you will be able to construct an amplifier using a pentode type tube, learn how to measure voltages and current

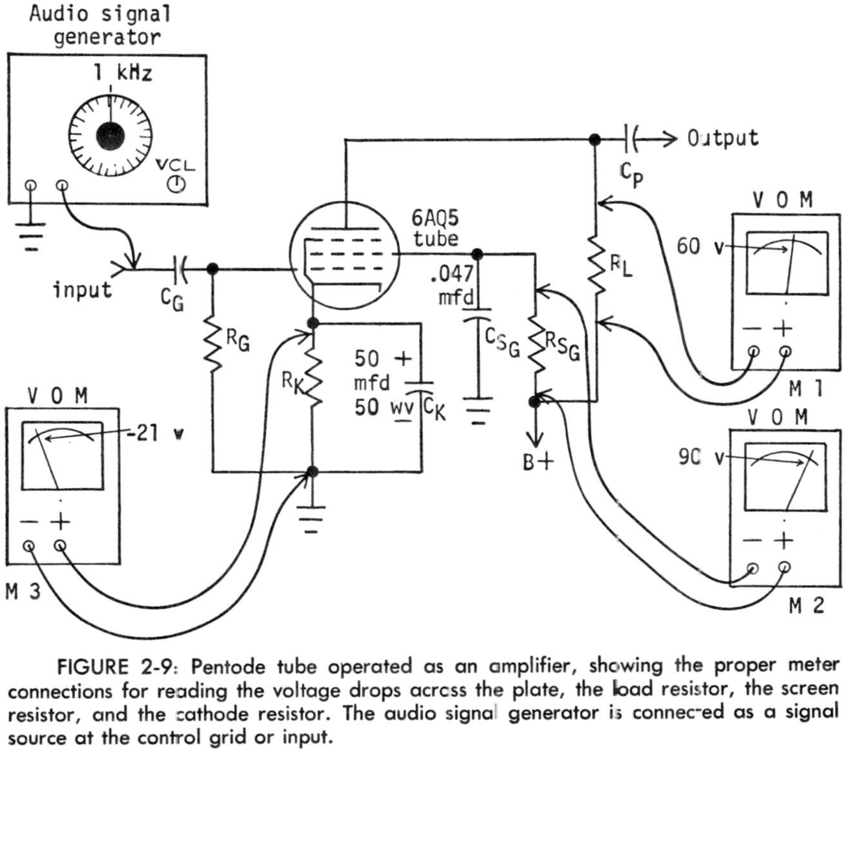

FIGURE 2-9: Pentode tube operated as an amplifier, showing the proper meter connections for reading the voltage drops across the plate, the load resistor, the screen resistor, and the cathode resistor. The audio signal generator is connected as a signal source at the control grid or input.

in this type circuit, and learn how to use the oscilloscope in checking this type circuit.

Refer to Figure 2-9. This illustration shows a 6AQ5 beam power pentode tube that uses a 9-pin miniature socket. Refer to your tube characteristic manual for element connections of the base. You will calculate the value of the resistors to be used in this circuit, and then use a VOM and check voltages and current to prove if your values are correct.

I should remind you before you start that all of the current flowing in the vacuum tube originates in the cathode circuit. On their way to the plate from the cathode the electrons are intercepted by the screen grid. Some are collected, but the majority go on to the plate.

When you are calculating the cathode current for a pentode tube, it will be screen current plus plate current ($I_{sg} + I_p$). When you look in the tube manual you will find the following characteristics for the 6AQ5 tube: plate voltage (E_p), 315 volts; plate current (I_p), 28ma.; screen grid voltage (E_{sg}), 250 volts; screen grid current (I_{sg}), 4ma.; control grid voltage (E_g), -21 volts. Assume that your supply voltage is (E_{bb}) 375 volts.

First, find the value of the plate load resistor, as follows: $E_{bb} - E_p = E_{RL}$; this is the amount of voltage the load resistor will drop ($375 - 315 = 60$ volts). The plate current will pass through this resistor and its value is 28ma., so use this formula: $R_L = \frac{E_{RL}}{I_p} = \frac{60}{.028} = 2142$ ohms. The nearest standard value is 2200 ohms. Now find the power dissipated by this resistor: $P = E \times I = 60 \times .028 = 1.68$ watts. The next highest value is 2 watts.

Now find the value of the screen dropping resistor: $E_{bb} - E_{sg} = E_{Rsg}$ ($375 - 250 = 125$ volts). This is the value the resistor will have to drop. $R_{sg} = \frac{E_{Rsg}}{I_{sg}} = \frac{125}{.004} = 3125$ ohms. The closest standard value is 3200 ohms. The power dissipated by this resistor would be: $P = E_{sg} \times I_{sg} = 125 \times .004 = .5$ watts. You can use a 1 watt.

Find the value for the cathode resistor. This is the screen grid current, plus the plate current: $I_K = I_{sg} + I_p = 28\text{ma.} + 4\text{ma.} = 32$ ma. The voltage dropped by this resistor is the grid bias, or 21 volts. $R_k = \frac{E_g}{I_k} = \frac{21}{.032} = 656$ ohms. The closest standard value is 650 ohms.

Find the power dissipated by this resistor: $P = E_{rk} \times I_k = 21 \times .032 = .672$ watts. Use a 1 watt.

Install these resistors in their respective locations as shown in Figure 2-9. This illustration shows a VOM connected across each of these resistors to measure the voltage drop across them. You can use one meter, and place it in each of the locations one at a time. M_1 placed across the plate load resistor should read 60 volts; this is the amount you calculated for the drop across this resistor. M_2 is shown across the screen grid resistor; this meter should read 125 volts—the value you calculated. M_3 is shown across the cathode resistor; this meter should read 21 volts, which is the grid bias for the tube.

While your meter is connected across this resistor, turn the audio signal generator on (shown at the input). Rotate the volume control slowly from minimum to maximum and watch the grid voltage change. This will simulate the voltage change during the input signal voltage. (We use this as an example only and not as a true reading.)

Connect the meter back to the screen grid as it was before. With the signal generator still connected to the input, rotate the volume of the signal up and down, to see if it has any effect on the screen voltage. It should not. This shows the effect of the screen decoupling capacitor. Disconnect this capacitor C_{sg}; again rotate the volume of the signal and watch the meter. The voltage does change some. Replace the capacitor while rotating the volume control; notice the voltage does not change after the capacitor is connected.

Move the meter back to the plate load resistor; again increase and decrease the volume control slowly on the audio generator. This is an example of how the instantaneous values of signal voltage will change the plate voltage to instantaneous values. You have witnessed how a tube amplifies.

Disconnect the meters, and connect them as shown in Figure 2-10. Switch your meters to read current. You will read current in the plate, screen grid, and cathode circuits. Observe the polarity of the meter leads. Connect them as shown in Figure 2-10. The current flows out of the plate circuit down through the load resistor. The negative lead should be connected to the plate, and the positive lead to the supply voltage.

M_1 is connected in the circuit as shown, and should read 28ma. or very close to this value. The audio signal generator should still be connected at the input. Rotate the volume control, slowly increasing and decreasing the input signal. Notice as you increase and decrease the input signal that the output will increase and decrease. This is an example of how the signal voltage changes the instantaneous values of bias voltage, which in turn changes the instantaneous values of the plate voltage and current.

Move the meter in series with the screen dropping resistor, shown

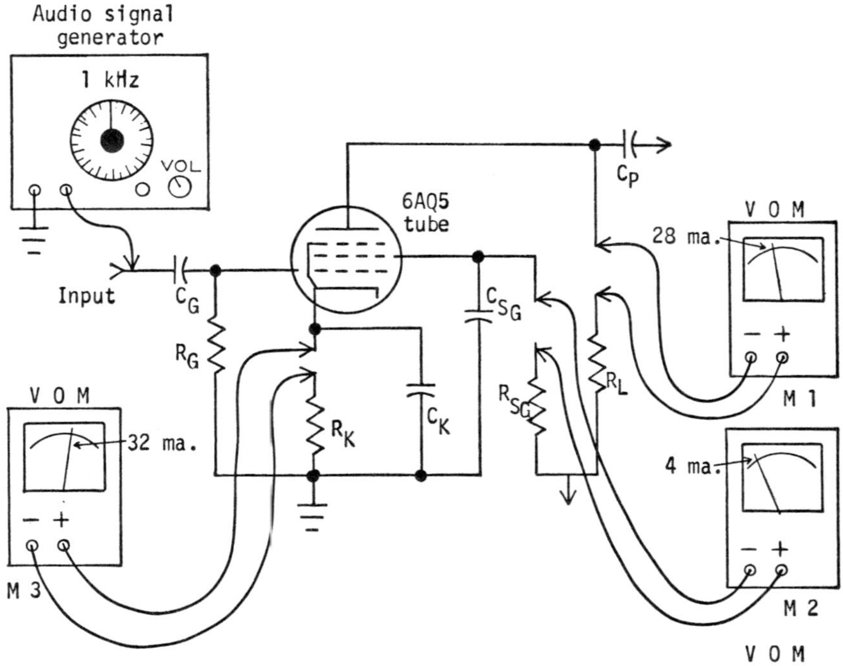

FIGURE 2-10: Pentode amplifier circuit showing the proper meter connections for reading the current in the plate circuit, the screen grid circuit, and the cathode circuit. An audio signal generator is connected as a signal source.

as M_2. This meter should read 4ma. or very close to it. This should make your calculations nearer to the resistor value you calculated.

Turn off the signal generator and move the meter down to the cathode circuit; install as shown (M_3). The meter should read the sum of the screen and plate current, which is 32ma. or the cathode current. Now turn on the signal generator, slowly increasing and decreasing the volume. Watch the meter reading as you do this; it should change as the input signal changes. You are viewing a changing input signal change the bias on the grid, which in turn varies the flow of current through the tube.

We have proven once again that grid bias controls the flow of current through the vacuum tube.

You will use the oscilloscope to check wave forms in a pentode tube amplifier.

Connect the oscilloscope as shown in Figure 2-11, at the top end of the plate load resistor. Turn on the signal generator and advance the

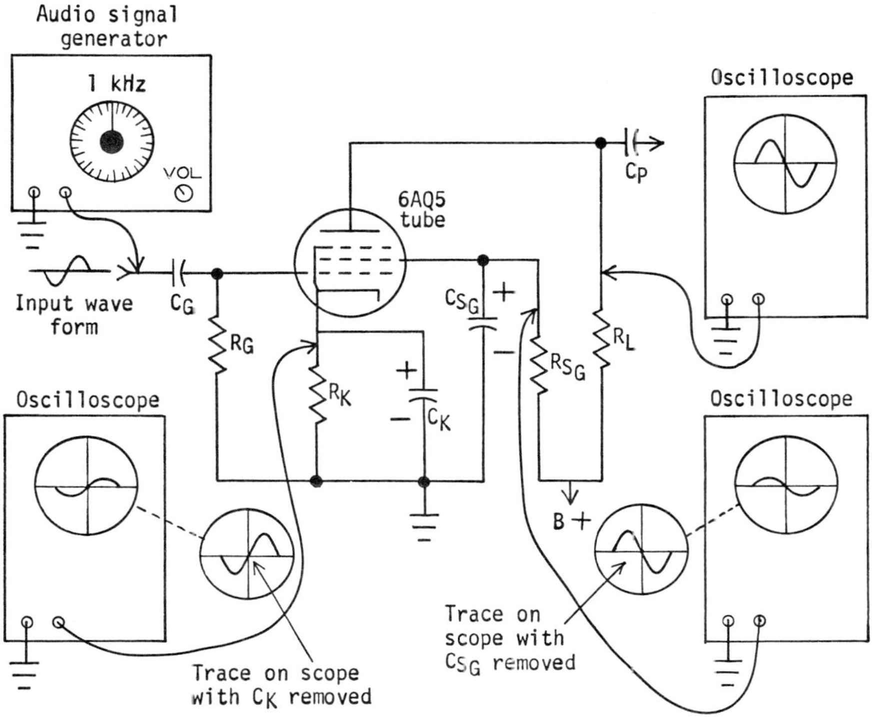

FIGURE 2-11: Pentode amplifier tube circuit showing an oscilloscope connected to show the wave forms at the various points in the amplifier circuit, and to prove the necessity for using some of the components.

volume control until a trace appears on the oscilloscope. Adjust the sweep on the oscilloscope to display 1 Hz of the 1kHz being applied. Increase and decrease the volume of the generator to simulate a changing signal at the input. The trace on the oscilloscope should increase and decrease. You are viewing this theory: as the input signal changes amplitude during each cycle, the output amplitude will change also.

Now change the oscilloscope lead to the top end of the screen resistor; notice you will have to turn up the vertical gain on the oscilloscope. Now turning the gain of the signal generator up and down has very little effect on the wave form of the oscilloscope. Turn the volume down on the generator, remove the screen decoupling capacitor, turn the gain up on the generator; notice you can see a large trace appear. This is the signal in the screen circuit. Touch the capacitor back on the screen; the trace should almost disappear. This shows the effect of the screen decoupling capacitor.

The traces shown in the illustration on the inside and outside of the oscilloscope are the wave forms you should observe during this test. While the capacitor is loose, place the oscilloscope lead back on the plate. Notice there is a loss in the gain of the output. Put the capacitor back on the screen while you have the lead on the plate. See how the gain in the output increases: this shows the overall gain of the tube is affected by this capacitor in the screen circuit.

You have observed with the aid of the oscilloscope the theory of the capacitor in the screen circuit.

Turn the gain down on the generator and move the oscilloscope lead down to the cathode circuit. Connect as shown; now increase the gain of the generator. You can see a slight trace on the oscilloscope; this proves the cathode capacitor is performing its function. You can prove this is true by removing the capacitor. See how the amplitude of the wave form increases; the capacitor must have been performing its function. Put the capacitor back while you have the signal going into the input and the oscilloscope still connected. Now you can faintly see the wave form, so again we have seen illustrated the theory of why the cathode capacitor is placed in the circuit.

HOW THE VACUUM TUBE AMPLIFIES THE SIGNAL

In the last four illustrations of both the triode and the pentode tubes we have used instruments attached to the various points of these tubes to measure both the static and the dynamic conditions. Static conditions were measured before the signal was applied, and dynamic measurements after the signal was applied. In the dynamic measurements, as you increased and decreased the volume on the generator, you noticed changes in both the voltage and current of the tube. In Figure 2-9 you noticed a change in the voltage drop across the load resistor in the plate circuit, and at the same time saw a change in the current passing down through this resistor. You also noticed a change in the voltage drop across the cathode resistor. What you were witnessing was amplification of a signal applied to the input and removed from the output, or a ratio between the voltage at the input and output. This we call voltage gain or amplification of voltage. It is caused by a changing voltage and current across a constant resistance or load, and can be expressed by the following:

$$\text{Voltage Gain (VG)} = \frac{\text{Voltage Out}}{\text{Voltage In}}$$

This formula refers to how much the voltage at the input has been increased at the output. You noticed as you were doing the previous experiments that voltage gain was measured across the plate load resistor.

When an AC signal is applied to a tube at the input, the grid and plate voltages will have fluctuating DC values. As the AC input sine wave changes alternations, going from zero to maximum in the positive direction, and changing direction and going from zero to maximum in the negative alternation, it will change the value of the DC voltage. This is an instantaneous value of DC voltage, caused by the variations in the AC signal. We can combine the AC signal voltage and the DC voltage algebraically, or the AC will add to or subtract from the DC value.

Refer to Figure 2-7(A) and mathematically calculate the voltage gain of this amplifier. This will explain how the tube amplifies.

In a mathematical formula, when referring to the static measurements, this is DC voltage and the symbol is a capital letter. When referring to the dynamic measurements, this is AC voltage, and the symbol is small letters. For example, if referring to the DC plate voltage, use the symbol E_p, and refer to plate current as I_p. These are the static symbols. When referring to the same element in the dynamic measurements, e_p (for plate voltage) and i_p (for plate current) would be the instantaneous values.

Use the following values: E_{bb} = 300 volts, E_p = 250 volts, I_p = 9ma., R_L = 5700 ohms, R_k = 900 ohms, E_g = −8 volts, signal voltage e_g = 4 volts P-P.

Refer to Figure 2-12(A). At the peak of the negative alternation of the input AC signal voltage sine wave, there will be the following conditions:

Instantaneous value of grid voltage:

$$E_g - e_g = e_c \qquad \begin{array}{r} -8 \\ +2 \\ \hline -6 \end{array} \text{ volts (operating bias)}$$

Instantaneous value of plate current:

$$i_p = \frac{e_c}{R_k} + I_p \quad i_p = \frac{6}{900} = .009 = .007 + .009 = .016 \text{ A}$$

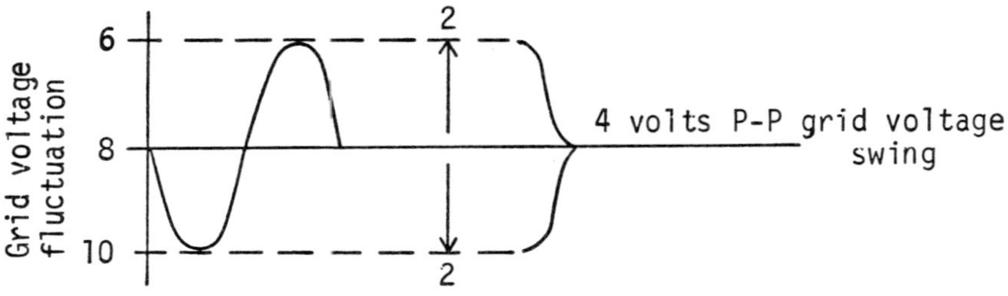

FIGURE 2-12(A): Grid voltage fluctuation caused by a variation of the AC signal input to the grid; the grid voltage swing is 4 volts P-P.

Instantaneous voltage drop across RL:

$$e_{RL} = i_p \times R_L = .016 \times 5700 = 91 \text{ volts}$$

Instantaneous value of plate voltage:

$$E_{bb} - e_{RL} = e_p \quad 300 - 91 = 209 \text{ volts}$$

This is a total change of 41 volts on the plate (250 − 209 = 41v), for a change of only 6 volts on the grid.

At the peak of the positive alternation of the input AC signal voltage sine wave, the following conditions will exist:

Instantaneous value of grid voltage:

$$E_g - e_g = e_c \qquad \begin{array}{r} -8 \\ -2 \\ \hline -10 \end{array} \text{ volts (operating bias)}$$

Instantaneous value of plate current:

$$i_p = \frac{e_c}{R_k} + I_p \quad i_p = \frac{10}{900} + .009 = .11 + .009 = .02 \text{ A}$$

Instantaneous value of plate voltage:

$$E_{bb} - e_{RL} = e_p = 300 - 11 = 289 \text{ volts}$$

Instantaneous value of voltage drop across RL:

$$e_{RL} = i_p \times R_l = .02 \times 5700 = 11 \text{ volts}$$

A total change of 39 volts on the plate (289 − 250 = 39 volts), for a change of only 10 volts on the grid.

Refer again to illustration 2-12(A). This shows that a 4-volt P-P AC signal voltage applied to the grid caused the grid voltage to fluctuate 2 volts more negative (here the voltage was going in a negative

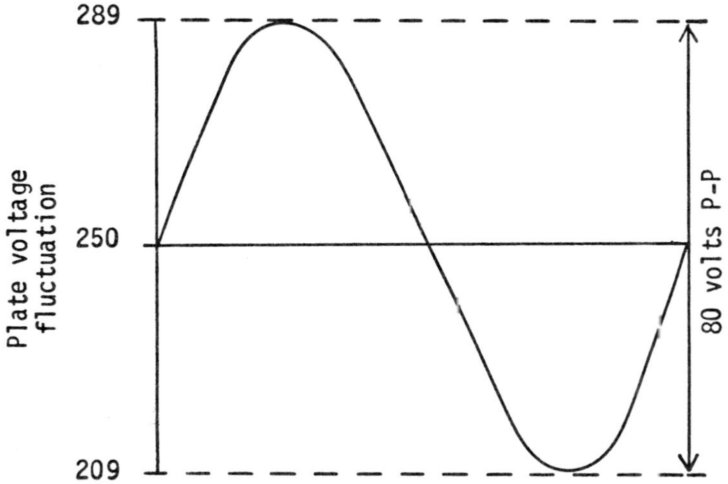

FIGURE 2-12(B): Plate voltage fluctuation caused by the variation of the grid voltage. 4 volts P-P at the grid has caused the plate voltage to vary 80 volts P-P.

direction) and 2 volts less negative (here the voltage was going in a positive direction). (B) illustrates that while the grid was going more negative it caused the plate voltage to increase; then it was going in a positive direction. While the grid voltage was going more positive, it caused the plate voltage to decrease.

From the formula given earlier in this chapter you can now find the gain for this amplifier.

$$VG = \frac{\text{voltage out}}{\text{voltage in}} = \frac{80}{4} = 20$$

The signal applied to the grid has been amplified 20 times, or this amplifier circuit has a gain of 20.

You have just proven another important theory in the amplification through a vacuum tube: the input signal is 180° out of phase with the output signal. This is true of any vacuum tube amplifier where the signal is applied to the grid and removed from the plate.

The illustration in Figure 2-12(A) shows the input signal; the first alternation is negative. The illustration in 2-12(B) shows the output signal; the first alternation is positive. In illustration (A) the second alternation is positive; in (B) the second alternation is negative. The complete input sine wave is 180° out of phase with the output.

LEARNING VACUUM TUBE SYMBOLS AND SOCKET BASE CONNECTIONS

There are so many different types and styles of vacuum tubes on the market today that identifying them as to their use becomes impossible. Every new piece of equipment presented on the market has new types or styles of tubes installed. The electronic technician, to keep abreast with this expanding need, should keep at his disposal the latest tube characteristic manual.

Almost all vacuum tubes can be identified as to the voltage used on the filament. In most cases, other than industrial and special purpose tubes, the first number of the tube indicates the amount of voltage used on the filament. For example, in the 6AX4 used as a damper in many of the television receivers, the first number indicates this to be a 6-volt filament tube. In the 3AU6 tube, this first number indicates the amount of voltage used on the filament. In the 17JM6A, the first

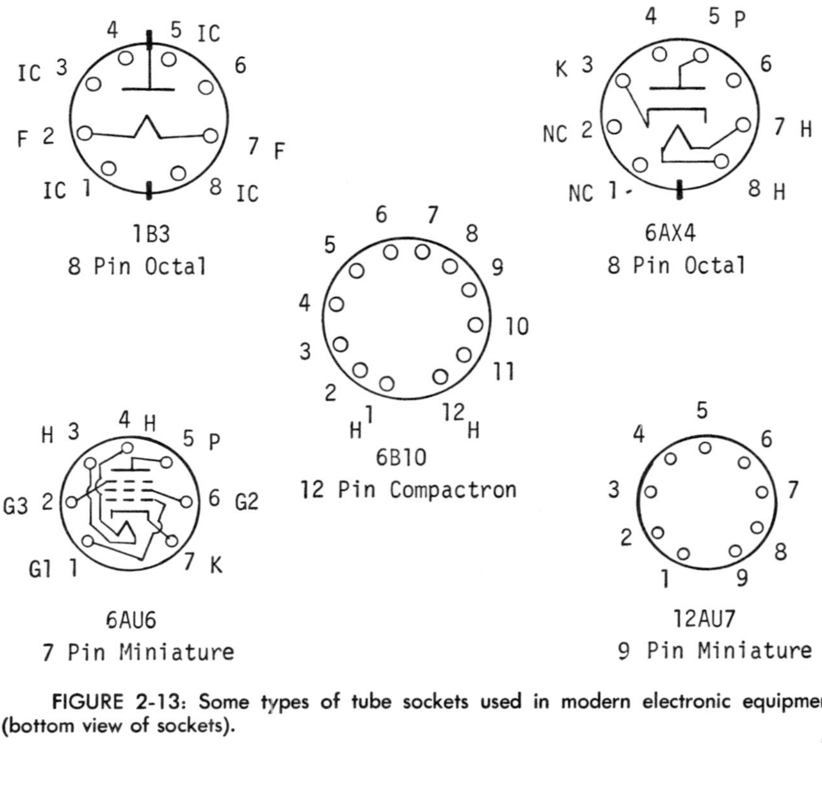

FIGURE 2-13: Some types of tube sockets used in modern electronic equipment (bottom view of sockets).

number indicates this is a 17-volt tube. After some experience you will be able to identify these tubes as to the filament voltage.

Figure 2-13 illustrates a few of the many tube sockets that may be found in electronic equipment. Each of these sockets shows the bottom view. The two sockets at the top of the illustration are eight-pin octal and the heavy dark line at the bottom of the circle represents a slot or hole in the socket. This is a guide to follow when installing the tube into the socket. All of the small circles are numbered and are used to identify the elements of the tube that is inserted into that position. For example: the grid of the tube may be connected to number 5 pin. To find the grid on the bottom of the socket, start counting at the guide pin or slot and count clockwise to pin 5. The two bottom sockets in the illustration are used for seven- and nine-pin miniature tubes. Notice at the bottom of the circle the spacing is wider between pins; this is a guide for placing the tube into the socket. To locate any element of the tube by number, with the opening closest to you start counting from the left side and count clockwise. The socket shown in the center is a special type used for the compactron type tube. The location of the elements is found the same way as that described for the octal socket.

SPECIAL TYPES OF VACUUM TUBES

There are many different types of special tubes and the scope of this book does not allow the coverage of every type. In Figure 2-14 are shown a few of the special types.

The symbol for a voltage regulator tube is shown in (a). This type of tube does not have a filament, and current flow is not by thermal emission. It is a cold cathode type tube; voltage applied to the plate causes the atoms to break down and become ions. Current flow through the tube is by ionization and the applied voltage will remain constant after ionization takes place. The potential placed on the plate differs from tube to tube, therefore these tubes can be used as voltage regulators. The current through the tube regulates the voltage regardless of the variation of voltage placed on the plate after ionization takes place.

Illustration (b) shows the symbol for a thyratron tube. This is usually a three-element gas tube. Current flow through the tube is by ionization; the tube acts very much like a triode before ionization. Plate current increases as plate voltage increases: plate current is controlled

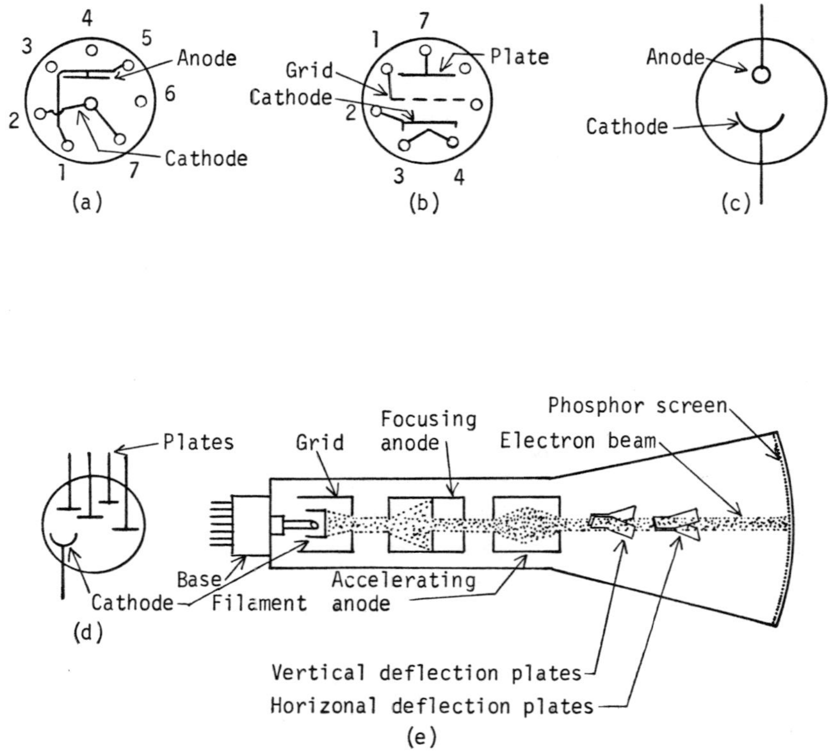

FIGURE 2-14: (a), (b), (c), and (d) illustrate the location of elements inside special tubes. (e) illustrates element location in electrostatic CRT.

by the voltage placed on the grid. When ionization takes place, space charge in the tube is overcome by the positive ions increasing the current and reducing the voltage drop in the tube. The positive ions are attracted by the negative grid and its effect is cancelled. When ionization occurs, the control grid loses control over plate current. This tube may be used as a relay where high current is used, and also has many applications in industry.

Shown in (c) is the symbol of a phototube. This tube is capable of converting light energy to electrical energy. Light striking a photosensitive material inside the tube causes electrons to be emitted. This type of emission is referred to as photoelectric emission. The photosensitive material can be made to respond to a variety of light frequencies, making it adaptable to many applications. One application for this tube is in the film projector to pick up the sound from the sound track of the film.

Illustration (d) shows the symbol of a photomultiplier tube. This tube is very similar to the phototube; however, notice it has more than one plate—it may have eight or ten plates. The cathode has a photosensitive material coated on its surface. This material has the ability to transform light energy into electrical energy when sufficient intensity of light strikes its surface. It emits electrons that are collected by positive plates. Each plate has a different potential applied. For instance, the first plate will have 100 volts applied; the second plate will have 200 volts. Each plate has 100 volts more potential than the last. Electrons are shifted from one plate to the other, and each plate has a higher potential than the last. The last plate is called the collector plate. Since each plate is capable of amplifying, it eliminates the need for several other stages of amplification. This tube may be used in the flying spot scanner, for both commercial and slow scanning.

Shown in (e) is the construction of the electrostatic cathode ray tube. The face of the glass envelope has a coating of phosphor on the inside. This material has the ability to change electrical energy to light energy. The filament heats the cathode as it does in other type tubes, and the cathode emits electrons. In order to form an electron beam focused on the face of the tube, a cylinder, called the grid, is placed around the cathode, as shown in the illustration. A negative potential is placed on this grid, forming an electrostatic field that repels the electrons. This aligns the beam of electrons and forces them through a small opening in the grid, called the aperture. The grid performs two functions: first it controls the number of electrons in the beam; second it concentrates them into a beam focused on the face of the tube. From the grid they enter the focusing anode, whose purpose is to focus the beam in a straight line. The beam then passes through the accelerating anode. Both this anode and the focus anode are at a positive potential. The electrons are attached and caused to accelerate along the path toward the face or screen. The beam then passes through two vertical plates called deflecting plates. One has a positive potential applied, and the other a negative potential. Here the beam is controlled by repulsion or attraction; the voltage applied is an AC voltage.

When the applied voltage is going through its positive alternation, one plate is positive and the beam is attracted in that direction. This causes the beam to sweep from the center to the top of the screen. When the applied voltage is going through its negative alternation, the beam is swept from the center to the bottom of the screen. The beam then passes on through the horizontal deflecting plates, and an AC voltage is applied to the horizontal plates. The beam is swept from left to right as the AC voltage passes through its cycle of alternation.

The phospor screen converts the signal applied to a light energy that is viewed on the screen as a wave form information. This type of tube is used in the oscilloscope.

Shown in Figure 2-15 is the construction of the electromagnetic deflection cathode ray tube. This tube is sometimes referred to as a CRT (cathode ray tube). The face of the tube is coated on the inside with a phosphor material, which has the ability to change an electrical energy to a light energy. The tube has a vertical and horizontal deflection coil placed around the neck of the tube on the outside, as shown in the illustration. There are two vertical and two horizontal coils. One horizontal coil is placed on the top and the other is placed on the bottom of the neck of the tube. One vertical coil is placed on the right side and one on the left side of the tube neck. These coils are fastened together so they form a ring around the neck of the tube on the outside (see the illustration). The two side coils, which are the vertical, are connected to the output of the vertical circuit from which they are fed a 60Hz AC pulse. This AC pulse causes a changing magnetic field to be built up around the two coils. The two horizontal coils are placed on the top and bottom of the tube neck. They are fed an AC pulse from the horizontal output at a frequency of 15,750 Hz, which causes a magnetic field to build up around them. Through this changing magnetic field the electron beam is passing, and by attraction and repulsion of these two changing magnetic fields at two different frequencies, the beam is caused to scan across the face of the tube, also from top to bottom of the screen. Electrical energy is changed to a light energy, and it is viewed as video information on the face of the tube.

There are many types, styles and uses of special purpose tubes. I would suggest that as you come in contact with a new type of tube, you refer to your tube characteristic manual and look up the specifications for the tube. There are different types of tube manuals: those listing the domestic type tubes (these are the ones we have been reviewing); those used in radio receivers and television; and some special types used in

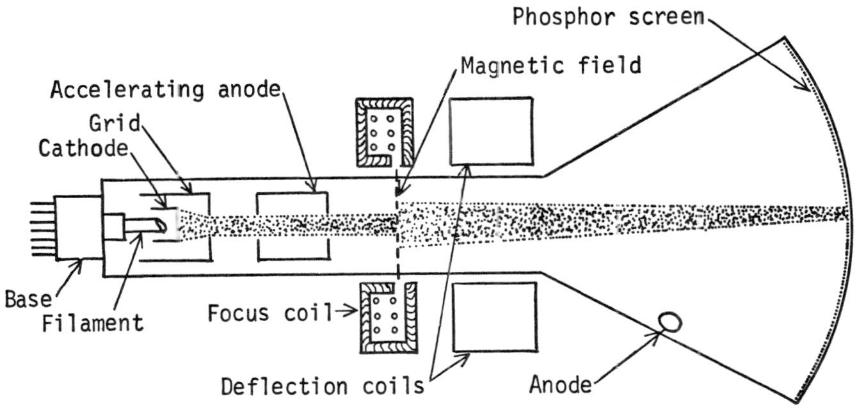

FIGURE 2-15: Construction of electro-magnetic deflection CRT.

other domestic equipment. There are manuals listing tubes used in industry and communications. There are also manuals listing all foreign type tubes and a cross-reference showing what type of USA standard tube might replace them. There are replacement manuals that may help you find a replacement tube when you cannot find the original. The regular tube manual can be very helpful when you do not have the diagram for the equipment you are working on.

3

Vacuum Tube Amplifier Circuits

VACUUM TUBE AMPLIFIERS CLASSIFIED ACCORDING TO FREQUENCY

An amplifier is a device that accepts a small signal at its input and increases it to a predetermined amplitude in its output. Since this type of amplifier is increasing the voltage, which we refer to as a signal, we could call it a voltage amplifier.

AUDIO AMPLIFIERS

If the signal voltage were coming from a microphone, a phonograph, or the second detector stage in a radio receiver, it would be considered an audio amplifier because the signal would be in the audio spectrum. The audio frequency extends from 15 Hz to 20KHz. An amplifier with a flat frequency response would be capable of producing this entire range and would be called a high fidelity amplifier. We sometimes speak in terms of an amplifier's quality. This refers to the ability of the amplifier to reproduce true fidelity or to produce a certain frequency range in its output without distortion. Usually an amplifier will consist of one or more voltage amplifiers in cascade. These are fed into an intermediate power amplifier or driver, which then feeds one or more power amplifiers in the final stage. Each stage is carefully matched, both in the input and output, so that the frequency will not be distorted by mismatching between stages.

RADIO FREQUENCY AMPLIFIERS

The radio frequency spectrum starts at 20 KHz and goes to approximately 300,000 MHz; an amplifier designed to amplify in this range of frequencies would be called an RF amplifier. Some RF amplifiers are designed to amplify only one frequency. For example, a crystal oscillator will amplify only one frequency—that of the crystal. RF amplifiers are designed with a tuned circuit on the input and output; this type will amplify a wider band of frequencies. For example, the RF stage in a radio receiver for the broadcast band is tunable from 550 KHz to 1600 KHz. Short wave receivers that use a switch to select the properly tuned circuit must amplify from 550 KHz to 30 MHz. In the FM radio receiver, the RF stage must tune a frequency ranging from 88 MHz to 108 MHz by using tuned circuits.

In communications, the radio transmitters, both for the broadcast band and for FM, use RF amplifiers for oscillators, buffers, frequency doublers and RF power amplifiers in the final stage. All of these operate at different frequencies, using tuned circuits.

An RF amplifier differs from the audio amplifier because it must operate over a wider band of frequencies. The RF amplifier can be operated as a class C amplifier where a large amount of power is required, but the audio amplifier cannot operate in this class because of the high distortion produced during amplification. The RF amplifier will pass the audio frequencies, but the audio amplifier may not pass the RF frequencies.

VIDEO AMPLIFIERS

Sometimes referred to as the wide-band amplifier, the video amplifier must amplify frequencies from 15 Hz to 4MHz all at the same time—for example, the IF amplifier in the TV receiver where the band width is 4 MHz wide. This amplifier, unlike the IF amplifier used in the AM radios that must pass only 10 KHz band width, must pass frequencies from 15 Hz to 4 MHz. This band width includes the audio and video frequencies and other frequencies which include the sync pulses, blanking pulses and the color information. The response of these amplifiers must be fairly flat in order not to cause distortion in the video information.

In order to extend the high frequency response in video amplifiers, peaking coils are used. At the higher frequencies the gain of an amplifier will start falling off due to stray capacitance formed between the tube socket and ground and between some of the components and ground. In order to eliminate the capacitance, peaking coils are placed in the circuit. Thus the inductive reactance cancels the capacitive reactance and the high frequencies are extended. This also aids in increasing the gain that is normally lost at higher frequencies. The use of IC chips which are frequency compensated in the modern TV receivers, aids in eliminating this high frequency loss as well as maintaining the gain.

IDENTIFYING COMPONENTS USED TO CONSTRUCT VACUUM TUBE AMPLIFIERS

The vacuum tube cannot amplify a signal alone; other components must be added to the circuit to aid the amplifier. In Chapter 2, we explained the importance of the loading in the plate circuit and how it was one of the active components in amplifying a signal. You also learned the different types of biasing and how it controlled plate current during amplification.

There are a variety of methods to transfer or couple the signal amplified in one stage to the next. Different methods will be discussed, identifying the parts and explaining the function of each.

If two amplifier stages are coupled together, a greater gain will be realized than from one stage alone. When coupling the two stages together, the output of one becomes the input of the next stage; this is called cascading.

AMPLIFIER CIRCUIT COUPLING

Capacitive and Resistive Coupling

Figure 3-1 illustrates capacitive and resistive coupling between two amplifier stages.

C_1 is the coupling capacitor; its purpose is to couple the amplified signal from the preceding stage to V_1. It also has a second important function: to block the DC voltage on the plate of the preceding stage from the grid of V_1.

R_1 is the grid resistor for V_1. Its purpose is to form a path to ground, or back to the cathode, for the excess electrons accumulated on the grid.

C_2 is the cathode decoupling capacitor. Its purpose is to bypass

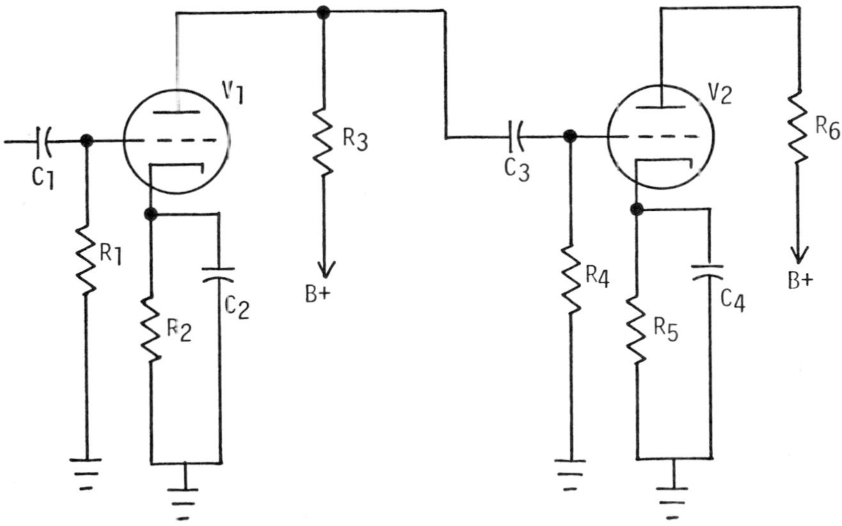

FIGURE 3-1: Capacitive and resistive coupling.

the AC signal around the bias power supply. Its second function, since a capacitor will oppose any change in voltage across it, is to act as a voltage regulator for the cathode bias supply. This capacitor is usually of the electrolytic type.

R_2 is the cathode resistor. The voltage drop across this resistor is the bias for the tube. This type of bias is sometimes referred to as automatic bias.

R_3 is the plate load resistor. The voltage drop across this resistor, subtracted from the supply voltage, determines the plate voltage on the tube. This resistor is usually made to be as high as possible and still stay within the limits of the tube rating. This factor helps to determine the gain of the tube.

C_3 is the coupling capacitor between the two stages. Its purpose is the same as that of C_1. However, it has another purpose, which is to help to determine the frequency response of this stage. This capacitor along with R_4 forms the capacitor-resistor network, and their values will determine at what frequency the stage will pass. The formula for determining this frequency is $f = \frac{1}{Rc}$. R is in megohms and C is in micro-farads. The values of these components are usually chosen for center of the frequency range.

R_4 is the grid resistor for V_2. Its purpose is the same as that of R_1. R_5, R_6, and C_4 act the same as R_2, R_3 and C_2.

Impedance Coupling

If you were to replace the load resistor in an amplifier stage with an audio choke, you would then be using impedance coupling. Since the DC resistance of the choke would be much lower than the plate load resistor normally used in the plate circuit, the voltage at the plate would be almost equal to the supply voltage. This would result in a greater output voltage or gain than possible with resistor loading. However, the impedance of the choke may not respond to all audio frequencies, and a frequency distortion would be introduced by using this type of coupling. A choke coil, if not properly shielded, would pick up stray magnetic fields and introduce a hum in the amplifier. Two amplifier stages using impedance coupling are shown in Figure 3-2. Other than the choke just described, the other components perform the same function as those in Figure 3-1.

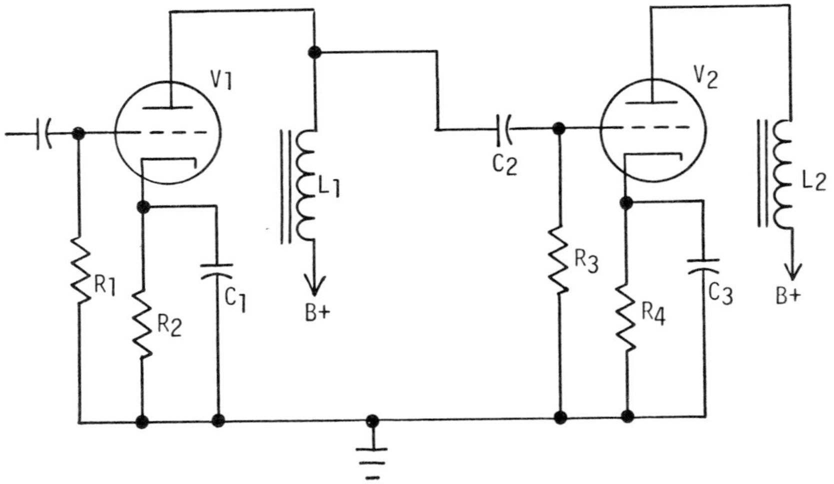

FIGURE 3-2: Two amplifier stages using impedance coupling.

Transformer Coupling

Figure 3-3 illustrates two stages using transformer coupling. All

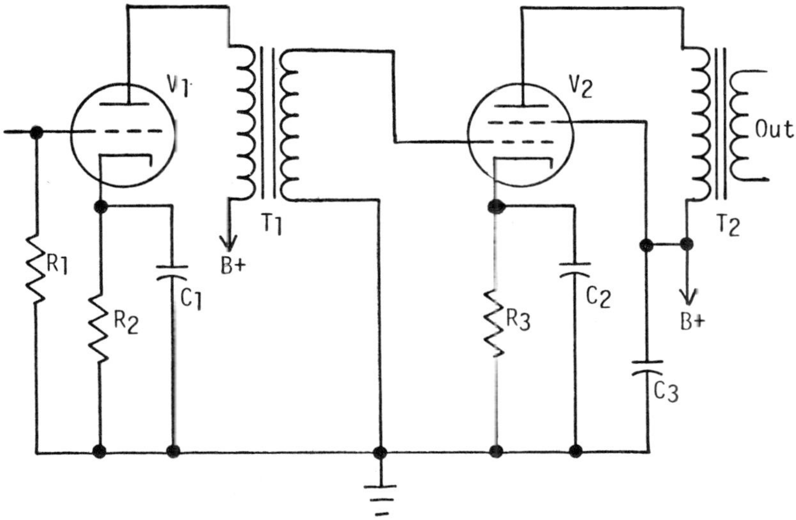

FIGURE 3-3: Two amplifier stages using transformer coupling.

of the components in this circuit, other than the transformer and the decoupling capacitor C_3, perform the same function as those in the previous circuits. The primary of T_1 is the plate load for V_1, while the secondary of this transformer acts as the grid return to ground, the same as the grid resistor in the previous circuits. Recall your theory concerning mutual induction. The current in one winding must be a varying current before we can have mutual induction. The signal at the output of V_1 causes the current in the primary of the transformer, which is the plate load, to vary. This varying current produces an EMF in the primary, which is induced into the secondary winding, which is then connected to the grid of the next stage. The transformer used to couple the two stages has made use of the theory that you learned earlier in electronics, namely, mutual induction, caused by a changing magnetic field. This transformer is usually the step-up type, which causes an increase in gain within the transformer itself. This is one of the advantages over some of the other types of coupling. Also, since it is possible to match the impedance of the input and output of this transformer, the amplifiers would have a better frequency response at certain frequencies.

C_3 is referred to as a decoupling capacitor; this means that since the screen grid is in the electron path in which the signal is being amplified, some of this signal is attracted to this grid, and the current flow through this grid is through the power supply. If this signal were allowed to return through the power supply, it would return to the cathode through the negative side of the supply. It would enter the electron flow once more, only now it would be out of phase with the signal present at this time. This would cancel part of the signal and reduce the gain in the tube. This would be referred to as degeneration or negative feedback. This capacitor provides a path around the supply for the AC signal to ground.

BASIC TROUBLESHOOTING VACUUM TUBE AMPLIFIER CIRCUITS

In Chapter 2 you were shown some of the instruments used to check a simple one-tube amplifier. You were shown how they should be connected into the circuit for checking various values of current, voltage, and resistance; also how to connect the audio signal generator to the circuit, to inject a signal; and how the oscilloscope should be connected to the circuit to check this signal. Some of these procedures will be repeated in learning how to test an amplifier with more than one stage. You can construct the amplifier shown. The values for each component is marked, or you may prefer to use one you have constructed. Whatever you choose to use, the testing procedure will be the same.

Refer to Figure 3-4. This illustrates a complete basic amplifier circuit. When illustrating methods for connecting test equipment, only that portion of the circuit will be shown. A power supply circuit is shown, but it will not be discussed at this time.

Assume that this amplifier is a phonograph amplifier or one that could be used with a tape player. The output of the crystal cartridge has been connected to the input, marked Signal In. A tape player head can be connected to the input, if you are going to use a tape player. If the amplifier has become inoperative, with no sound coming from the speaker, check the tubes first. Approximately 90% of the trouble in tube type amplifiers is tube failure.

If you find the tubes check all right (no bad ones), you now have a choice of two methods that can be used to locate the trouble:

1. The voltage-resistance check, starting at the output and measuring through to the input.
2. The signal injection method, starting at the input or the output.

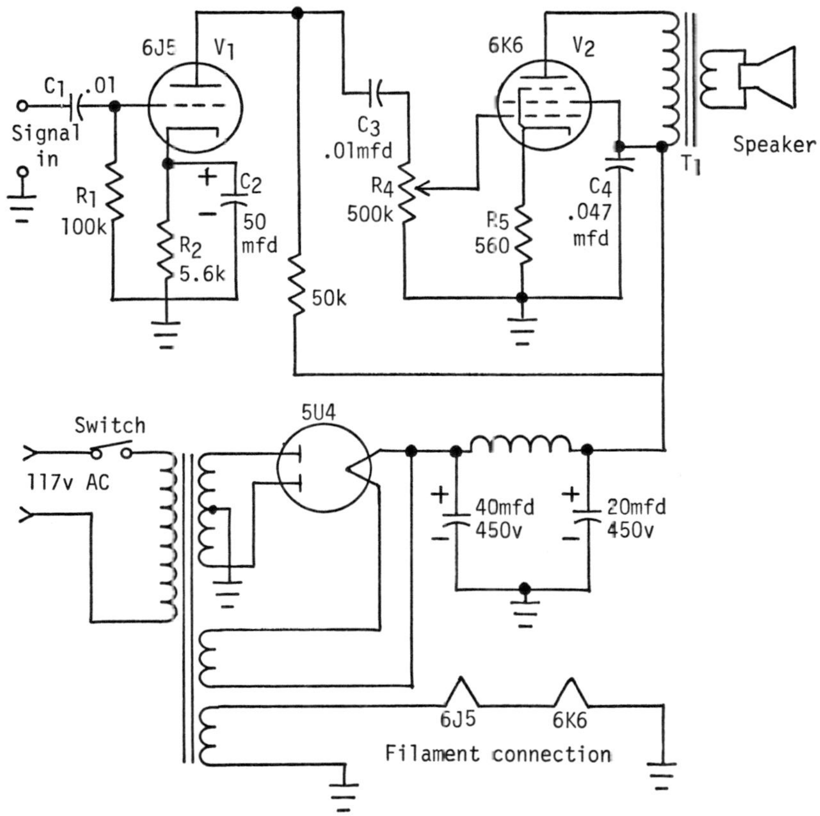

FIGURE 3-4: Two-tube audio amplifier and power supply.

This time you will begin by using the voltage-resistance test and start at the input.

There is never a high potential of voltage at the input side of C_1, and since it is not likely to be defective, go to R_1. There is never a high current flowing in the grid circuit; therefore it is not likely that this resistor is defective. The next step to check would be the grid bias. If you find the bias correct, there would be no use checking R_2 or C_2 because a defect in either of these components would affect the bias.

Figure 3-5(A) shows how a voltmeter should be connected to measure the cathode bias. If you find this bias measures correctly go to the plate to measure the plate voltage. Figure 3-5(B) shows how the voltmeter should be connected to measure the plate voltage. If the plate

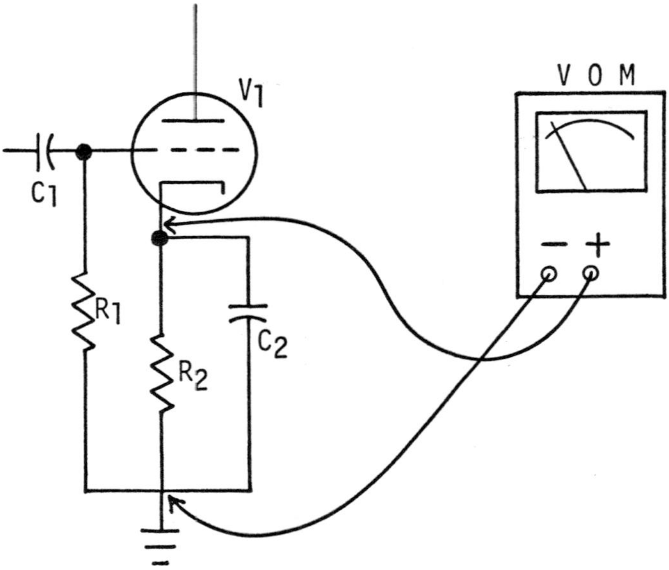

FIGURE 3-5(A): How the voltmeter is connected to read the grid voltage.

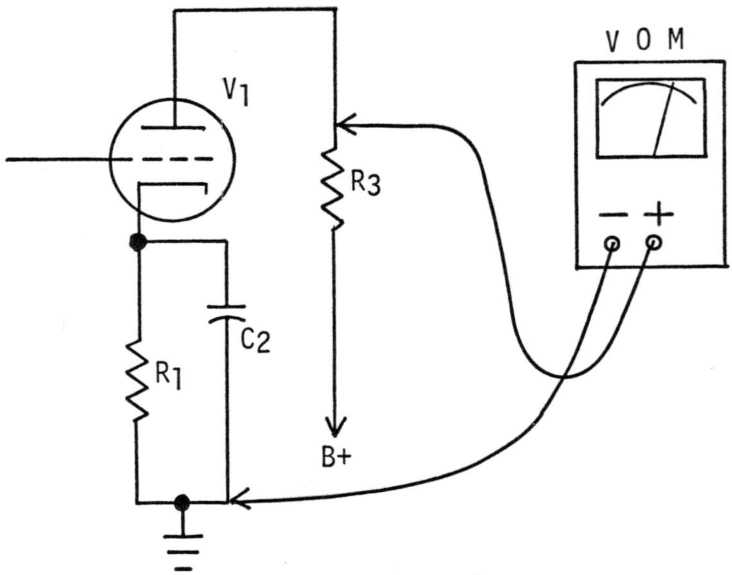

FIGURE 3-5(B): How to connect the voltmeter to measure the plate voltage.

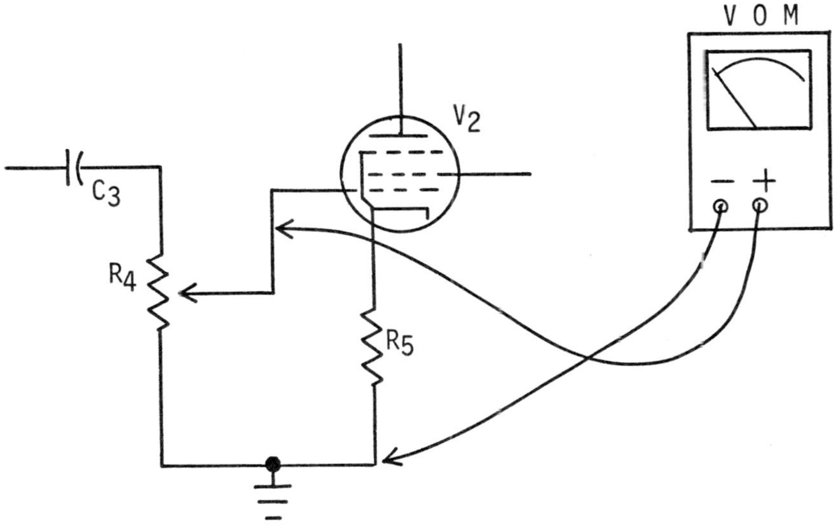

FIGURE 3-6: How voltmeter should be connected to V_2 grid for possible positive voltage.

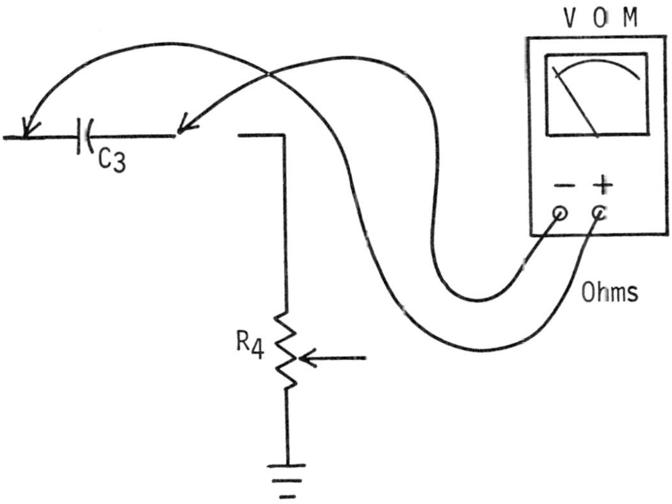

FIGURE 3-7: How the ohmmeter should be connected across C_3 after it has been removed from the circuit, to determine if it may be open.

voltage is correct, go to C_3. There is a high positive potential on one side of this capacitor and a low potential on the other side. As there is a possibility this capacitor is shorted, measure from grid of the next stage tube to ground; if the capacitor were shorted, there would be a positive voltage on the grid. Figure 3-6 shows how the meter should be connected to measure for positive voltage on the grid. If the capacitor is not shorted, it could be open. Turn the power off and switch the meter to read resistance on the highest scale. You will check the capacitor for possible open. Figure 3-7 shows how the meter should be connected across the capacitor after it has been removed from the circuit.

Carefully check this capacitor with an ohmmeter, set on the highest range on the ohms scale. Connect the ground lead of the meter to the ground side of the capacitor; then watch the meter closely and touch the positive side of the meter to the positive side of the capacitor. If the capacitor is good, you will see only a slight upward movement of the pointer and then it will quickly fall back to infinity. If the capacitor is open, the pointer will not move. If the capacitor is good, reconnect it into the circuit and reset the VOM to read voltage. Move to V_2 and check the grid voltage. The meter does not read, indicating no grid voltage. Since the grid voltage is developed across the cathode resistor, turn off the power and disconnect the ground end of the resistor. This step is not necessary in this case but would be required when a cathode

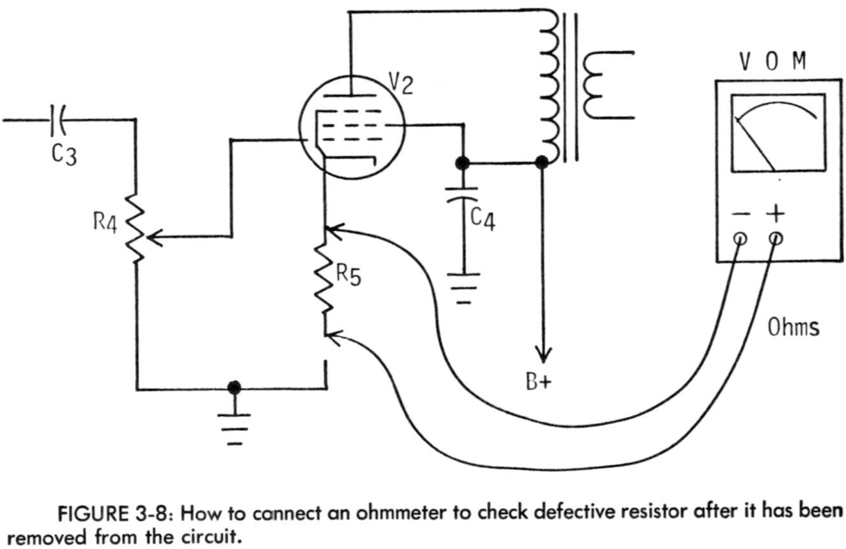

FIGURE 3-8: How to connect an ohmmeter to check defective resistor after it has been removed from the circuit.

bypass capacitor is used. In case there should be a capacitor across the resistor, we have shown you how it should be done.

Figure 3-8 shows the proper method of connecting the meter across the resistor after VOM is switched to the resistance scale. The meter is indicating the resistor is open. Replace the resistor and the amplifier is back in operation.

4

The Intriguing Oscillator Circuit

HOW THE OSCILLATOR CIRCUIT OPERATES

Since to most electronic students, the oscillator circuit seems difficult to understand, perhaps a definition of the word *oscillator* should be given at this time. We will approach the oscillator circuit in this manner: first, a definition of the word; then a definition of what the circuit does; and last, how the circuit performs and where it performs, along with troubleshooting hints.

Webster's Dictionary defines *to oscillate* thus: to swing to and fro; to fluctuate; to cycle or vibrate.

In physics it would mean to vary between maximum and minimum values as an electric current or a periodic change in a variable, as in the amplitude of an alternating current.

The Electronic Dictionary defines oscillator as "a circuit that generates alternating current at a frequency determined by the values of its components." The oscillations are produced by using positive feedback with an electron tube, transistor, magnetic amplifier, or other amplifying device. Oscillations are maintained by a power source, and frequency stability may include a crystal, a tuning fork, or other unvarying sources of vibrations.

In the study of AC analysis, you learned that the electronic oscillator was one of two methods for generating a sine wave. The oscillator circuit can produce two types of wave forms, the sinusoidal and the nonsinusoidal. The sinusoidal is the sine wave and the circuit producing it is called the sinusoidal oscillator. The nonsinusoidal is any wave form other than the sine wave, and the circuit producing it is called a nonsinusoidal oscillator.

A term used in our definition of an oscillator was feedback; this refers to feeding back or transferring a voltage from the output to the input. During amplification through any tube or transistor, the input is out of phase with the output by 180°. Since the oscillator depends upon a feedback voltage for continued operation, this voltage is fed back in phase. When a voltage is fed back, that is in phase with the input, this is called positive feedback, or regeneration. When a voltage is fed back that is not in phase, it would cancel the input voltage and the oscillator would cease to function. This type of feedback is called negative feedback or degeneration.

TYPES OF OSCILLATOR CIRCUITS

Oscillator circuits are divided into two types, audio and radio frequency, or AF and RF. The main difference between the two circuits is the frequency at which the oscillation takes place.

One very important thing for you to remember is that the vacuum tube or the transistor does not oscillate. The definition stated that *the frequency of oscillation is determined by the components in the circuit.* The vacuum tube or the transistor merely amplifies the oscillating voltage produced by these components when power is applied.

Audio Oscillators

This is a circuit in which the oscillating frequency is between 15 Hz and 20 KHz, in the audio frequency range. The fundamental wave form produced is the sine wave; therefore, it is an AC voltage. The components used in this circuit are usually an RC network, or resistor and capacitor. Some circuits such as the vertical and horizontal oscillators in the television receivers use the LC or inductance and capacitance. The RC feedback network, in an oscillator circuit from the output back to the input, shifts the phase of the voltage until it is in phase with the input. When an AC voltage is applied to a resistor and capacitor in series, the voltage across each component is out of phase with the applied voltage. The phase difference will depend upon the frequency of the applied voltage, the value of the resistor and capacitor. Since the output is 180° out of phase with the input, if we were to cascade three RC networks and shift the phase across each one by 60°, the voltage would be in phase at the input. This is positive feedback or regeneration to continue oscillation in the circuit as long as power is applied.

Figure 4-1 illustrates an audio phase shift circuit, using three RC networks coupled in cascade. Because the signal at the output of this amplifier is 180° out of phase with the signal at the input, and to cause this amplifier to oscillate, an output voltage should be coupled back to the input that will be in phase with this input voltage. To accomplish this three different RC networks are coupled as shown, and each network shifts the voltage by 60°. The voltage arriving at the input will be in phase with the input voltage and will add to it, causing the circuit to oscillate and continue oscillating until the supply voltage is removed.

Figure 4-2 illustrates an oscillator circuit using a transformer for regenerative feedback. This circuit is used in the vertical oscillator circuit in some older models of television receivers. The frequency is 60 Hz and is determined by the time constant which is $C_1 \times (R_2 + R_3)$.

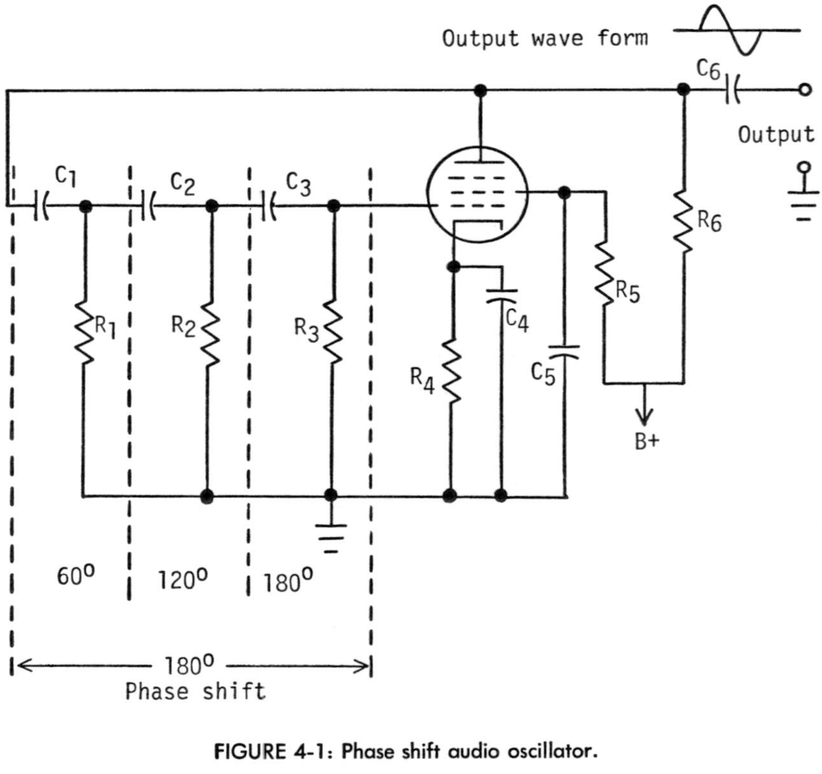

FIGURE 4-1: Phase shift audio oscillator.

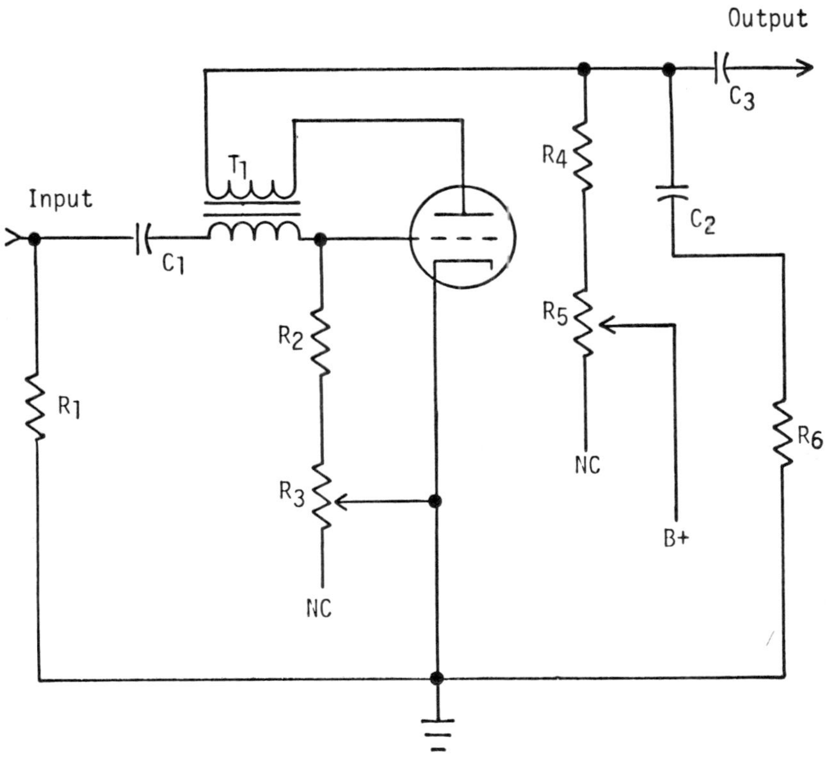

FIGURE 4-2: An oscillator circuit using a transformer for regenerative feedback. This circuit is called a blocking oscillator.

R_3 is the frequency control. Increasing its value will decrease the frequency, and decreasing its value will increase the frequency. In the formula T = RC, if we change either the value of R or C, T will change. This control, R_3, would normally be the vertical hold control on a television receiver that prevents the picture from rolling up or down. This voltage, generated by C_1, R_2, and R_3, is applied to the grid of the tube and amplified. The primary of the transformer is the plate load. This voltage at the plate of the tube is 180° out of phase with that at the input. As the current passes through the primary winding of the transformer, it is changing (because it is an AC current), and this changing magnetic field produces an EMF. When this EMF builds up to its maximum value, it induces, by mutual induction, a voltage into the secondary winding. This voltage now is 180° out of phase with that of the primary voltage. We have shifted the output voltage back in phase with that of the input. Since this voltage is in phase it will add to the input voltage and oscillation will continue.

The series resistors R_4 and R_5, which are also a part of the plate load, affect the amplification of the tube, which increases or decreases the height of the picture. Thus R_5 would be the height control on the television receiver. The C_2 and R_6 network is called the wave-shaping network. They change the wave form into a spiked wave form which is coupled to the vertical amplifier through C_3. The wave form produced by this circuit is nonsinusoidal or sawtooth. The network just described shapes it to the proper timing so that it can be applied to the vertical deflection coils.

Since the frequency of this circuit is 60 Hz, which is in the audio range, it is considered in the audio oscillator class.

Radio Frequency Oscillators

This is an oscillator circuit that normally uses a tuned circuit in which the frequency of oscillation is generated. This frequency begins at 20 KHz and includes the long wave, medium and short wave, VHF, UHF, micro-wave, and goes into the gigahertz frequency.

Radio frequency oscillators (RF osc.) are divided into two types: self-excited or shock-excited, and crystal control. A self-excited oscillator depends upon its own resonant circuit for initiation of oscillation and frequency determination. The self-excited oscillator is usually made up of a coil and a capacitor or an LC circuit involving what is called the flywheel effect or the oscillation of electrons back and forth, generating an AC voltage. This, you will recall, you learned in your study of current flow in a parallel resonant circuit. Current will continue to flow in the circuit after the source has been removed. Figure 4-3 illustrates a parallel LC resonant circuit. If we were to close the switch momentarily and then open it, we would simulate the oscillation that is generated in an LC circuit. The instant the switch is closed, the capacitor will charge up to the source voltage and current will flow in the circuit from the negative side of the capacitor up through the coil to the positive side until the capacitor is completely charged. Shown in (a) by the short arrow, only a small amount of current will flow at this time. This, if you recall, is due to the inductance of the coil opposing the flow of current through it. In (b), if we open the switch, the

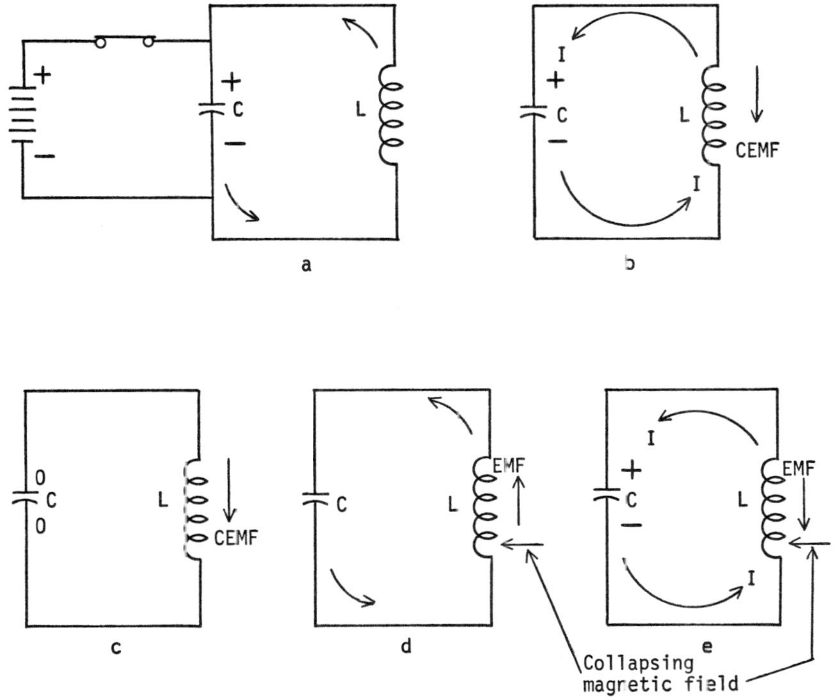

FIGURE 4-3. Basic LC circuit showing the different stages during the oscillating cycle.

capacitor will start discharging and current will start flowing up through the coil, creating a magnetic field and a CEMF (counter electromotive force). This CEMF opposes the discharging of the capacitor, but the source is greater than the opposition and the capacitor continues to discharge. As current continues to flow up through the coil, the magnetic field becomes stronger and the CEMF increases; (c) shows a point now is reached where the CEMF equals the source. The voltage now is zero in the circuit and the current will cease to flow. Since the magnetic field in the coil now has no source to keep it expanded, it starts collapsing back into the coil. Circuit (d) shows the field entering back into the coil, which creates an EMF (electromotive force) in the circuit opposite to that of (b). This force now starts current flowing in the circuit and the capacitor starts charging back to its original polarity, as shown in (e). One cycle of AC has been completed, but due to resistance of the coil, this condition will continue for only a short time before it is dampened out. By using a positive feedback and supplying a new source of energy at the completion of each cycle, we can cause the circuit to continue to oscillate. The charging and discharging of a capacitor and the expanding and contracting of the magnetic field about an inductor have generated an oscillation which is the same in all types of LC oscillator circuits.

Figure 4-4(A) illustrates the Hartley series-fed oscillator, and Figure 4-4(B) the Hartley shunt or parallel-fed circuit. Both are widely

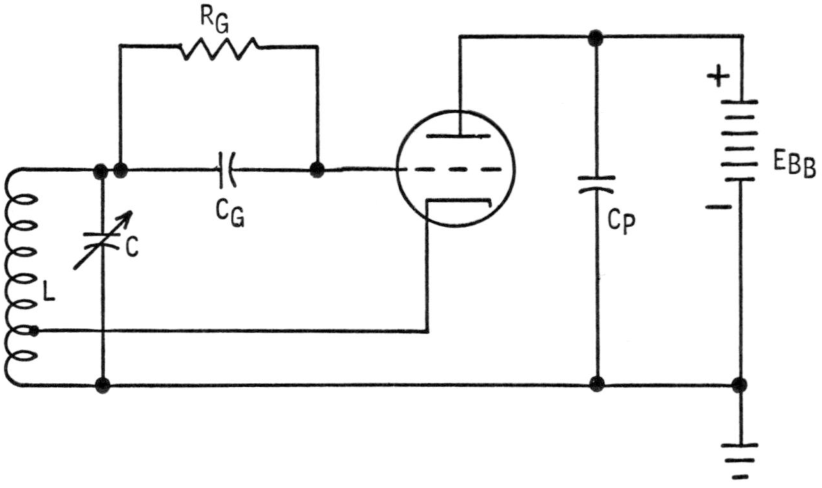

FIGURE 4-4(A): Hartley oscillator series-fed.

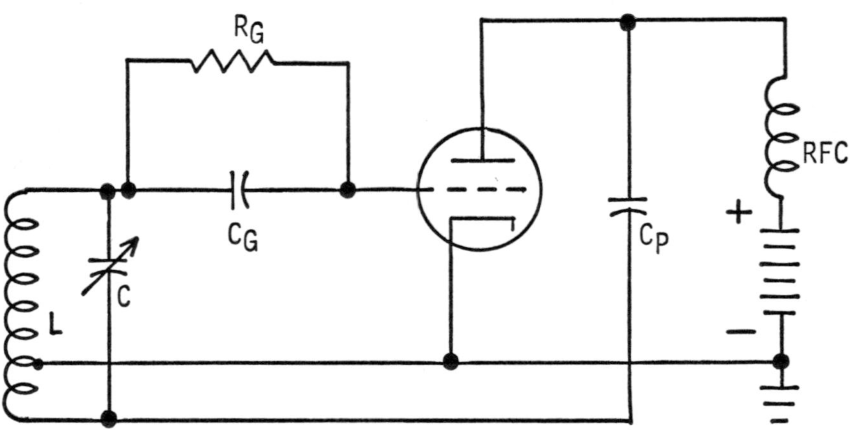

FIGURE 4-4(B): Hartley oscillator shunt-fed.

used in heterodyne circuits in radios, TV, and FM receivers. The difference between these two circuits is that in the series-fed, the feedback network (which is the bottom half of the coil) is in series with the power supply. In the parallel-fed, the bottom half of the coil (which is the feedback network) is in parallel with the power supply. In both circuits, the tap on the coil separates the input and the output circuits. From the tap to the grid is the input LC, which generates the frequency. Current passing up through the output coil produces a magnetic field that induces a voltage into the input coil, which is then in phase, and oscillation is sustained.

The ECO, or electron coupled oscillator, is one of the most widely used variable frequency oscillator circuits. Most oscillator circuits, like the one just described, will shift frequency with any sudden change in their load. The use of a pentode tube in this circuit makes possible greater amplification plus the isolating of the oscillator stage from the amplifier and external circuits. The circuit is very similar to the Hartley circuit, with the screen acting as the anode. Oscillation takes place on the grid, cathode, and screen circuits. Grid voltage is developed during the time the circuit is oscillating, which will vary the plate current at the same frequency and cause the tube to amplify. The only connection between the plate circuit and the oscillator circuit is through the electron flow between the screen grid and the plate. Any change in the plate supply or circuit components will have very little effect on the oscillation frequency. Where a more stable frequency is desired, we usually refer to the crystal oscillator circuit.

Crystal Oscillator Circuits

In oscillator circuits, such as commercial transmitters, where frequency stability is a factor, crystals are used to generate the frequency. The crystal oscillator is not subject to drift with changing plate voltage or other circuit components as is the LC circuit. The crystal is a substance called quartz; it looks very similar to a piece of frosted glass and is mined in its natural state. It is cut into thin slices and then ground to the desired frequency. The frequency generated by the crystal depends upon its size and thickness. If it is to operate at high frequencies, it is ground very thin; for low frequencies it is much thicker.

The crystal is suspended between two metal plates and placed in an enclosure called the holders. If an AC voltage is applied between the two plates, it will cause the crystal to distort its shape in one direction and then reverse itself, distorting in the opposite direction. It would go through a mechanical vibration and a small EMF would be produced during each reversal of this vibration. This process of converting an electrical energy to mechanical energy and then back to electrical energy is known as *the piezoelectric effect.*

Even though the frequency of oscillation is fairly stable in the crystal oscillator circuit, some precautions are taken to insure a more accurate stability:

1. A well-regulated power supply will prevent plate voltage variations.
2. The crystal is placed in a temperature regulated oven and a constant temperature is maintained.

Figure 4-5 shows a basic crystal oscillator circuit. Notice there is no coupling between the crystal and the grid; the crystal will block any DC flow between the grid and cathode. Notice too, the input stage is not tuned; the frequency is preset by the crystal itself.

Figure 4-6 represents a Pierce crystal oscillator. One advantage of

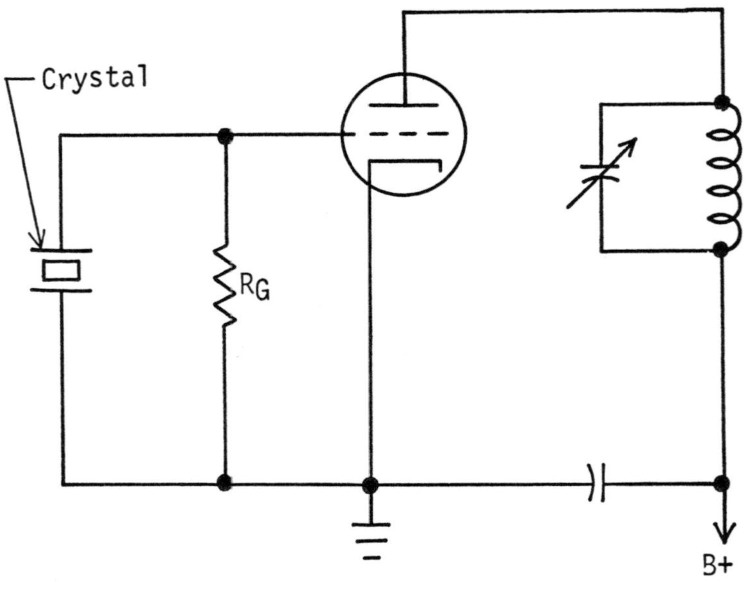

FIGURE 4-5: Crystal oscillator circuit.

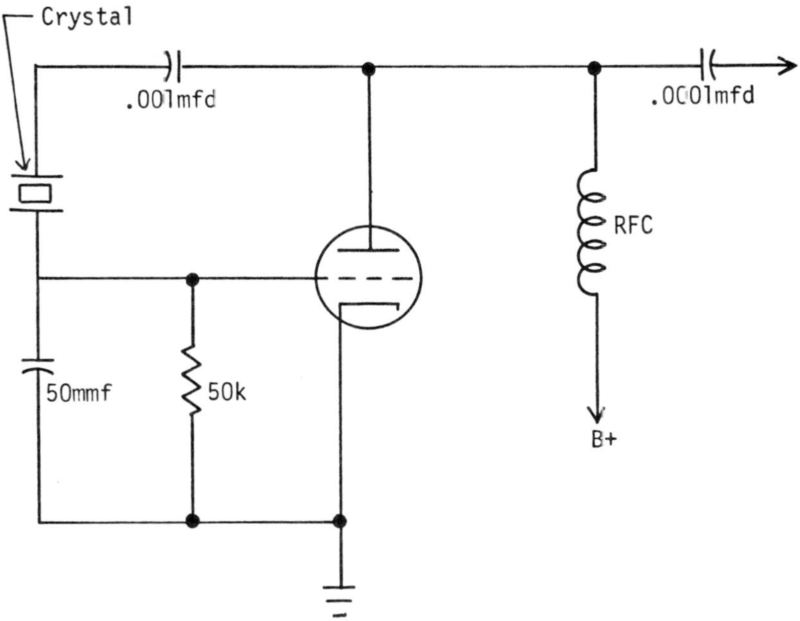

FIGURE 4-6: Pierce crystal oscillator circuit.

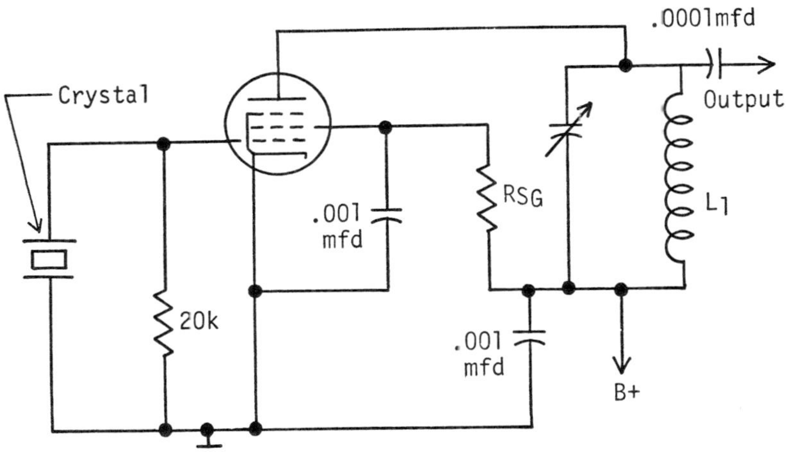

FIGURE 4-7: Pentode crystal oscillator circuit.

this circuit is that it does not require any tuned circuit. The plate potential must be kept fairly low in this circuit, which limits the output power, and a capacitor is placed in series with the plate and the crystal to prevent any DC voltage strain across the crystal.

Figure 4-7 shows a pentode crystal oscillator circuit. Even though the interelectrode capacitance is reduced in the pentode tube over the triode, by the use of the extra elements (sustained oscillation is dependent upon grid-to-plate feedback in the tuned plate/tuned grid circuit), the gain of the pentode tube is so great and the Q of a crystal is so high that the small capacitance formed between the grid and plate in the pentode tube will feed back enough voltage to sustain oscillation. This circuit will produce a greater power output than with the triode tube, without damage to the crystal.

THE OSCILLATOR'S PLACE IN AN ELECTRONIC CIRCUIT

Without the oscillator circuit, it would be impossible to have any form of satisfactory radio or television communication. Along with the many other electronic circuits that depend upon the oscillator for their operation, the RF carrier of the radio, of FM, and of television transmitters is produced by the oscillator stage. Without the carrier, the intelligence to be transmitted could not be sent out. Without the oscillator circuit, the heterodyne circuit could not function in the radio or television receiver. Most test equipment would not be available without the use of the oscillator circuits. None of the timing or switching circuits, including the computer, would operate without the use of an oscillator circuit of some type. This should give you some idea of why you should understand the oscillator circuit. It is probably the most important circuit in the electronic field today. In the study of any oscillator circuit, you must consider three important requirements:

1. You must have a resonant circuit, with components capable of producing a certain frequency when excited or when power is applied, that will start a fluctuating motion or vibration.
2. The small voltage produced in the oscillator circuit must be amplified through a vacuum tube or transistor.
3. You must have some form of feedback in order to keep the oscillation set-up in the resonant circuit continuing. This should be a positive feedback or regeneration method, so that

the feedback voltage will be in phase and add to the input voltage.

VACUUM TUBE AND TRANSISTOR OSCILLATOR CIRCUITS

We have covered only a few of the many oscillator circuits. However, in Chapter 11, the multivibrator oscillator is covered. We have not discussed those oscillator circuits that might be considered special because they will only operate at certain frequencies. These are the ones in the UHF and SHF (super high frequencies) frequencies.

Figure 4-8 illustrates a UHF TPTG (tuned plate/tuned grid) quarter-wave length hairpin tank oscillator. This circuit oscillates in the 300-400 MHz band. It resembles that found in the UHF tuner of a television receiver. In the modern receivers this is a part of the integrated tuner circuit, but the principle of operation is the same. The circuit is called the hairpin because the grid and the plate coils are small pieces of wire, bent into a U shape, which resemble the old-fashioned hairpin. These hairpin tank circuits, along with the variable capacitors, form the grid and plate resonant circuits and are coupled to the common cathode.

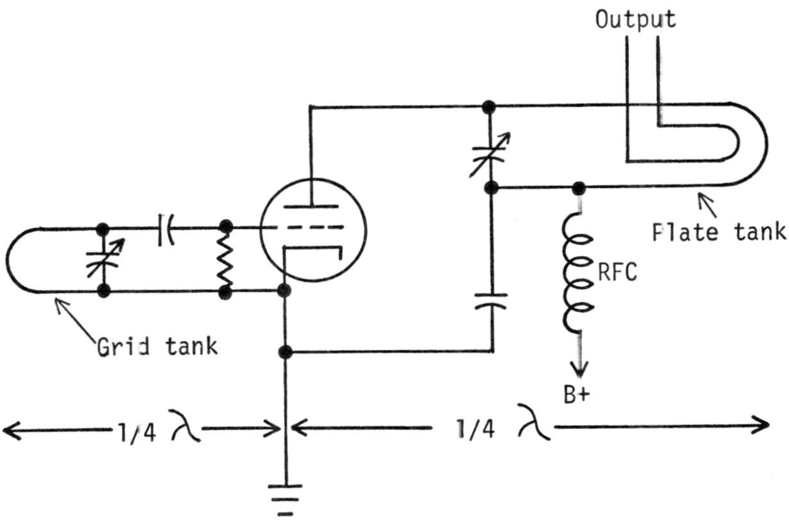

FIGURE 4-8: UHF TPTG quarter-wave length hairpin tank oscillator circuit.

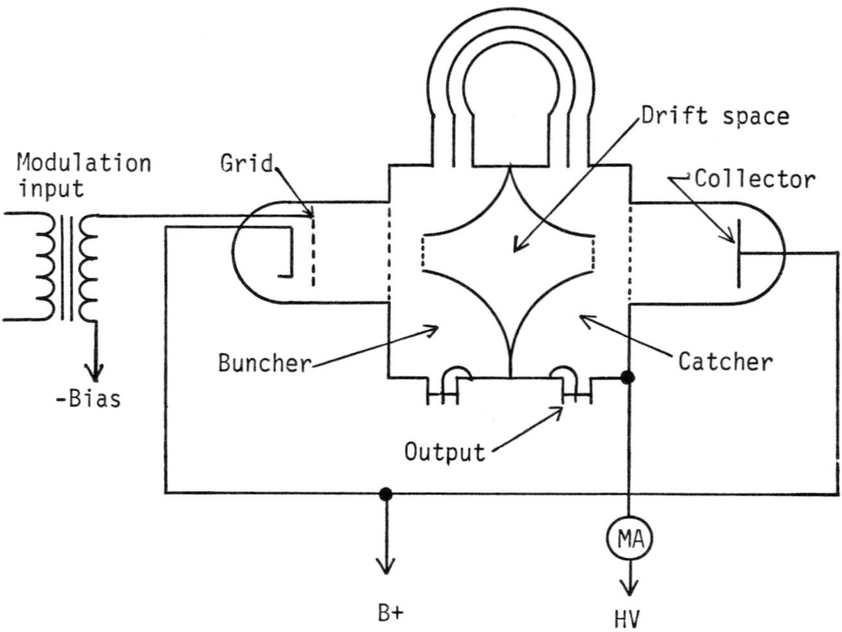

FIGURE 4-9: Klystron oscillator circuit showing the feedback loop coupling the frequency controlling cavities.

Figure 4-9 is another special type of oscillator circuit. It is the Klystron oscillator used in the micro-wave frequencies. Electrons from the cathode pass through an electric field created by two grids, in a cavity called the buncher. The resulting effects here are slow and fast moving electrons "bunched" or formed into groups, which forms a velocity-modulated beam. Next, the beam travels to a catcher cavity again, passing through two parallel grids; here the beam produces an RF voltage between the grids. The catcher cavity is made resonant at the velocity-modulated beam; frequency of oscillation is generated by the passing of the beam through the grid aperture. A feedback loop is provided between the two cavities and oscillation is sustained.

Figure 4-10 illustrates a solid-state RLC oscillator circuit, using a SCS (silicon controlled switch). When the switch is closed, capacitor C charges through inductor L to a voltage above the supply, if its Q is high enough. When the current reverses, the diode will not conduct and the SCS is triggered. This will start the capacitor discharging, which will cause the anode gate to go to near a ground potential. Now

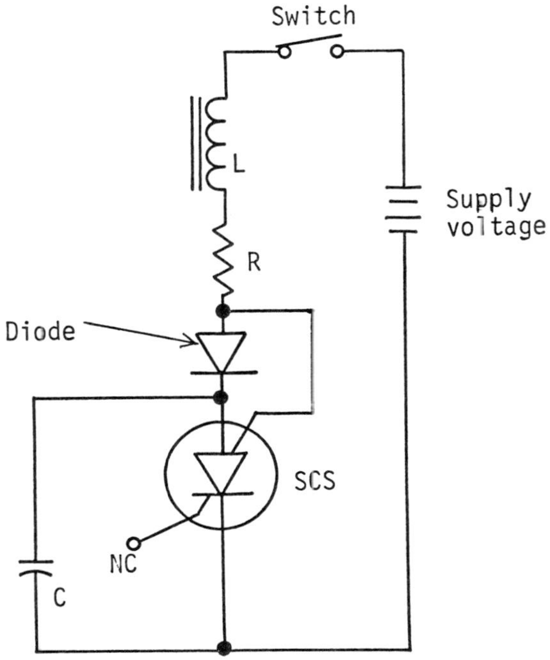

FIGURE 4-10: RLC silicon controlled switch oscillator circuit.

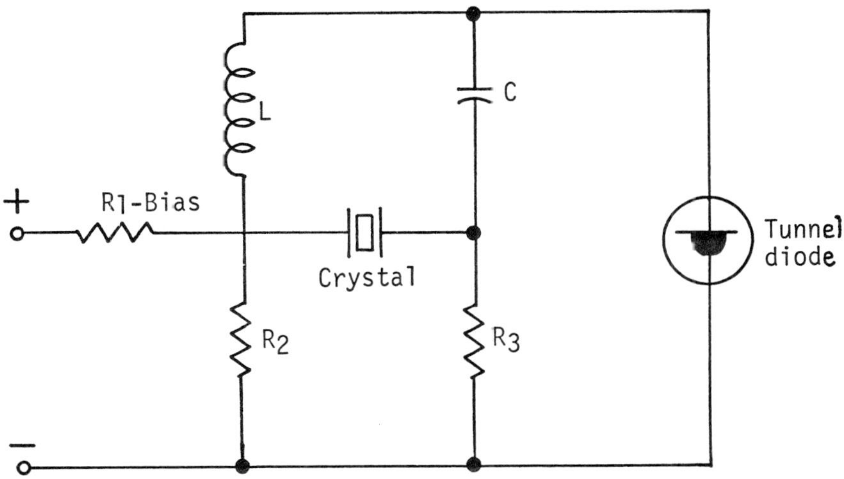

FIGURE 4-11: Crystal controlled oscillator circuit using tunnel diode.

the anode will not have sufficient holding current, and the SCS will turn off; then the capacitor will start charging again, to begin another cycle.

Figure 4-11 illustrates another solid-state type of oscillator circuit—crystal controlled, using a tunnel diode. This circuit operates similarly to the tube type crystal oscillator, frequency being controlled by the type of crystal used. This circuit is used in some of the modern transmitters operating in the SW (short wave) band from 27 MHz to 50 MHz, very low power and short range.

METHODS USED TO TROUBLESHOOT OSCILLATOR CIRCUITS

We will discuss two methods that can be used to troubleshoot oscillator circuits, using the VTVM and oscilloscope. You have learned that the bias for amplification in an oscillator circuit is produced by the oscillation generated by the components and the feedback network. In a vacuum tube, this voltage is produced between the grid and the cathode. Therefore, to measure this voltage, the test instrument must be connected between these elements. Figure 4-12 shows how the instrument should be connected. Either the VTVM or a digital volt-

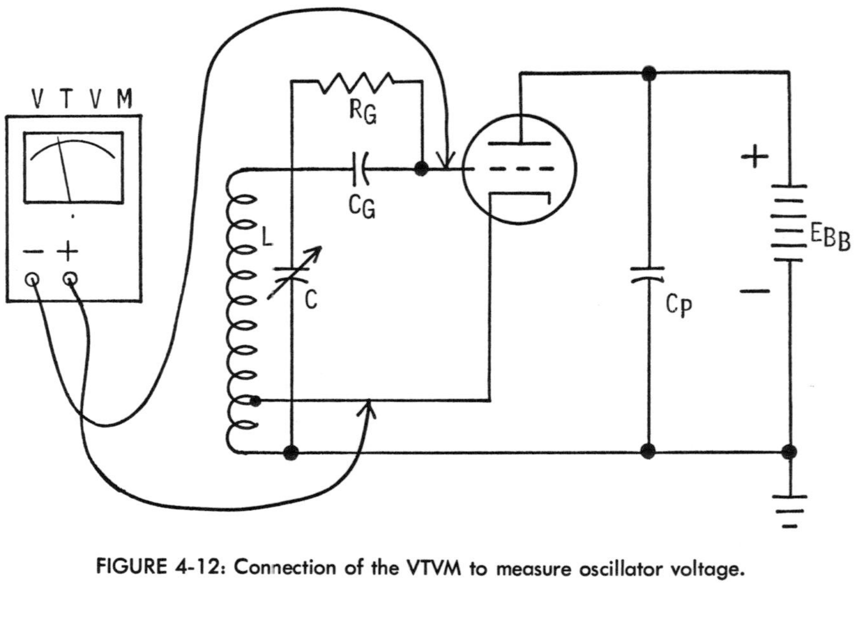

FIGURE 4-12: Connection of the VTVM to measure oscillator voltage.

ohm meter must be used to make this test. The oscillator circuit is the most critical circuit of all electronic circuits; the slightest change in the circuit will cause a shift in the frequency. The instrument used should have as much isolation from the circuit as possible. This circuit shows the meter is connected to the grid and the cathode. The polarity of the leads means that the negative lead is connected to the grid and the positive lead is connected to the cathode. When measuring voltage between the control grid and the cathode, the grid is negative and the cathode is positive.

If you were checking a piece of electronic equipment in which the complete operation was dependent upon the oscillator circuit, this would be one of the methods you would use to make sure the oscillator was operating. If your reading on the meter was the same or near that value recommended for normal operation, you would know the oscillator was performing. The circuit shown is one that is used in many of the superheterodyne receivers instead of the triode tube. The circuit would be connected to the oscillator section of a pentigrid-converter tube. If you were checking the oscillator of a radio receiver, the meter would be connected as shown. To prove that this reading is the oscillator voltage, with the meter connected as shown, rotate the tuning capacitor from one end of the dial to the other. The value should change as you go higher or lower in frequency; if it does not change, you are not reading the oscillator voltage. VTVM can be used on any circuit to measure oscillator voltage regardless of the type of components used or the type of circuit.

A method for connecting the oscilloscope to an oscillator circuit is shown in Figure 4-13. The oscillator is generating an RF voltage, and the oscilloscope controls will have to be set at a high speed rate. The display will depend upon the frequency of oscillation. If the frequency is low, the display may show only a few cycles, but as the frequency increases, the number of cycles will increase. We can use the radio receiver to illustrate this. Connect the oscilloscope lead as we have shown, and rotate the variable capacitor to the low frequency end of the dial, or 550 KHz. The display should look somewhat like that shown in (a) of the illustration. Now rotate the dial toward the high end of the dial, 1600 KHz; notice as you go higher in frequency how the number of cycles are increasing. When you are at the high end, your display should look close to that of (b). This method will hold true for any type of oscillator circuit in which the frequency can be varied. In a later chapter we will show how these same methods can be applied to the transistor circuits.

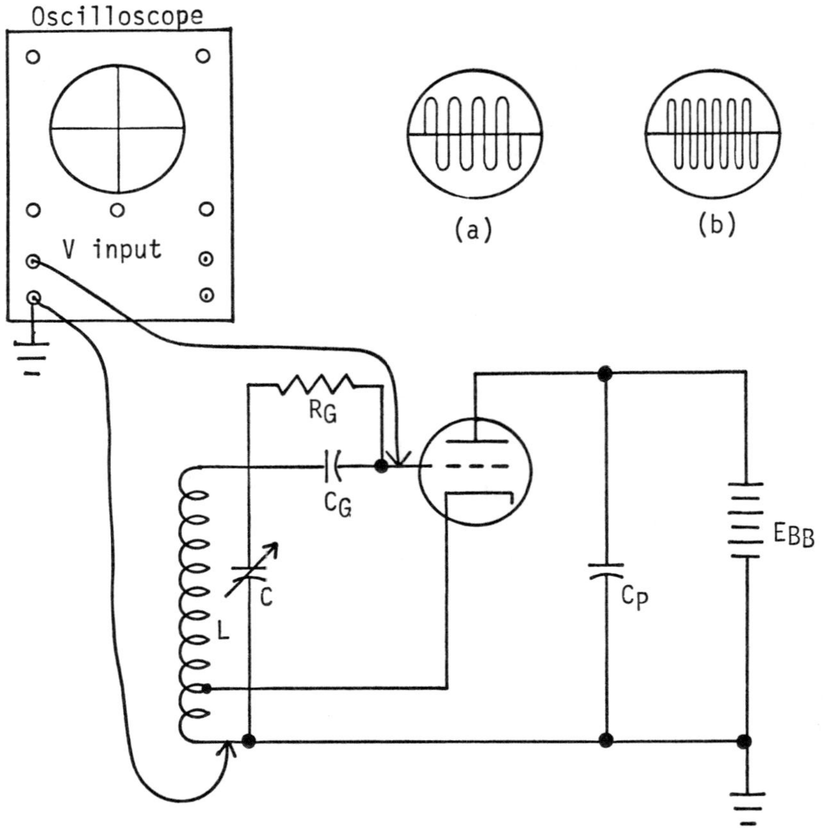

FIGURE 4-13: Illustrating how the oscilloscope is connected to the oscillator circuit to test its performance. (a) and (b) show possible wave forms.

5

The Power Supply as an Energy Source

A power supply is defined as a power line, generator, battery, power pack, or other source of power for electronic equipment. Energy is defined as ability to do work. We can say now that the source that causes electronic equipment to do work, or operate, is the power supply. The description of a power supply just given is applied to electricity and electronics, but if we confine our definition to electronics only, then the meaning is changed slightly. For example, the power supply for your home is the power line, but the power supply for your radio or television is not the power line; these devices have their own power supply. Here the difference is the voltage; the power line furnishes AC voltage, while the power supply for the radio or television is DC voltage.

The energy source for all electronic equipment is a DC power supply, from the digital wristwatch which uses a miniature battery, to the largest computer using a large DC power supply. The large DC power supply receives its energy source from the power line; this is the type of power supply to be described in this chapter. The electronic equipment receives its energy from the power supply, and the power supply receives its energy from the power line.

RECTIFICATION: HALF-WAVE AND FULL-WAVE

Rectification is defined as the process of converting an alternating current to an undirectional current. This means changing AC voltage to pulsating DC voltage, and then eliminating the pulses of DC voltage by the use of the correct filtering network.

Rectification is performed by two methods, half-wave and full-wave.

Half-Wave

Refer to Figure 5-1. Here is the power line connected to the primary or input of a power transformer. The power transformer is connected to a diode vacuum tube, and a resistor is connected between ground and the cathode of the diode tube. The resistor represents a load connected to the output of the tube; this is only an example to show how rectification is performed.

The power transformer is a step-up and a step-down type of transformer. One of the secondary windings has more turns than the primary, therefore it steps the voltage up. One of the second windings has fewer turns than the primary, therefore it will step the voltage down; this is why it is called a power transformer. The voltage induced into the secondary is 180° out of phase with that of the primary and at the input or primary of the transformer is shown one Hz of the 60 Hz AC voltage. The alternation above the reference line is positive voltage, and the alternation below the reference line is negative voltage. Current will pass through the tube only when the anode or plate is positive and the cathode is negative.

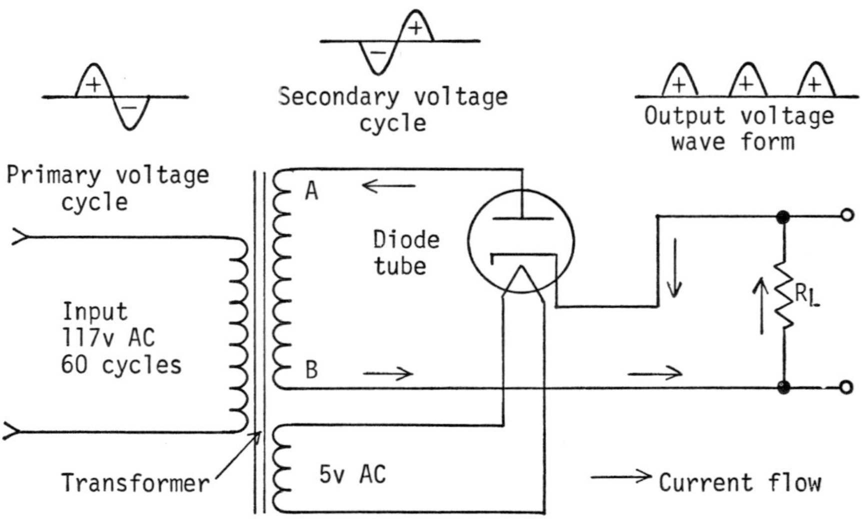

FIGURE 5-1: Basic half-wave rectification.

One Hz of the 60 Hz is shown above the primary of the transformer; one Hz of the 60 Hz is shown above the secondary of the transformer, reversed. In order to understand this action better we will divide this cycle into alternations. The first alternation is negative; this makes point A at the top of the transformer negative and the bottom at point B positive. The condition now is that the plate is negative; therefore the cathode is positive and no current will flow. This is indicated by the first blank space on the output voltage wave form. During the positive alternation of the secondary voltage cycle, point A at the top of the transformer is positive, and the bottom at point B is negative. Current flows from point B to the resistor and up through it, producing a voltage drop across this resistor, to the cathode up through the tube to the plate and back to the transformer, secondary. The current flow through the resistor has produced a voltage in the output as shown by the positive alternation at the output wave form. Rectification was accomplished by undirectional current flow through the tube and current flow through the tube's load resistor. Since rectification took place only during the positive alternations of the secondary voltage cycles, only half of the wave form was produced in the output, hence half-wave rectification. Since only half of the wave form was produced, the frequency of the output voltage is the same as the output voltage, or 60 Hz.

Full-Wave

As the name implies, both halves of the wave form are rectified, and the output frequency is twice that of the input, or 120 Hz. This type of circuit uses a different type of rectifier tube, called a duo-diode. It has two diodes built into one glass bulb. Refer to Figure 5-2 for a basic full-wave rectifier circuit, and notice that the transformer has a center tape. This is necessary in full-wave rectification. One Hz of the 60 Hz is shown at the input: the secondary is lettered A, the center is lettered B, and the bottom is lettered C.

The wave form for the first alternation is shown at the top of the secondary, marked as it would appear for the first alternation. As the AC in the secondary winding changes direction, the wave form shown above the secondary winding will reverse also.

We shall discuss this process one alternation at a time, as we did the half-wave circuit.

During the first alternation, the following conditions will exist: A

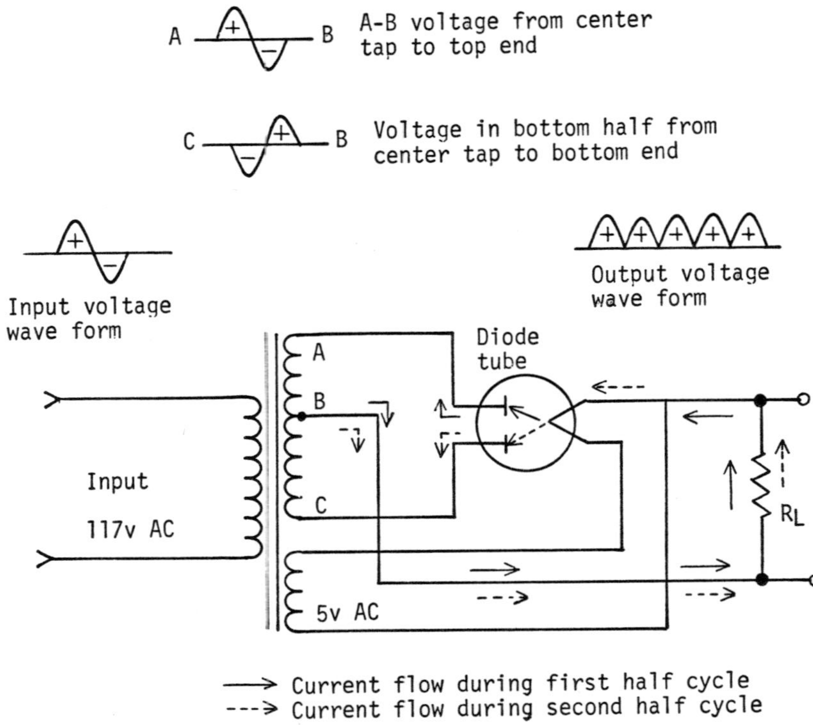

FIGURE 5-2: Basic full-wave rectification.

at the top will be positive, B at the center will be negative with respect to A, the bottom at point C will be negative, and point B will be positive with respect to point C. Follow the solid arrows, which indicate the current flow during the first alternation. Notice current flowing from the center tap toward the resistor, up through the resistor to the tube. One plate is connected to the positive end of the transformer, and the other plate is connected to the negative end. In order for the tube to conduct, the plate must be positive with respect to a negative cathode; this is the condition at this time. The current flows up through the tube to the top end of the transformer and back to the center tape from where it started. The current would not flow across the other plate of the tube, for it is connected to the negative end of the transformer. Now the AC reverses and the condition is that the top end of the transformer at A is negative and point B is positive with respect to point A, and the bottom end at point C is positive and point B is negative with respect to point

C. Refer to the two wave forms at the top of the secondary winding and reverse them for this description. The dotted lines show the current flow for this half of the cycle: current flows out of the center tap to the resistor, up through the resistor toward the tube. The cathode is negative with respect to the plate that is connected to the bottom end of the transformer or point C, which is positive. Current will pass in this direction to the bottom of the transformer and back to the center tap from where it started. Current will not flow through the other plate, as it is negative at this time. We have completed one cycle and produced a voltage in the output for both halves of the cycle. As current passes up through the resistor during each half-cycle, it causes a voltage drop across this resistor and produces an output as shown by the output wave form. Notice that the voltage wave forms are all positive, because the tube would only conduct when it was connected to a positive half-cycle. There are other types of both half-wave and full-wave rectifiers; we shall discuss one other type of half-wave.

Transformerless Half-Wave Rectifier

This power supply does not use a transformer between the power

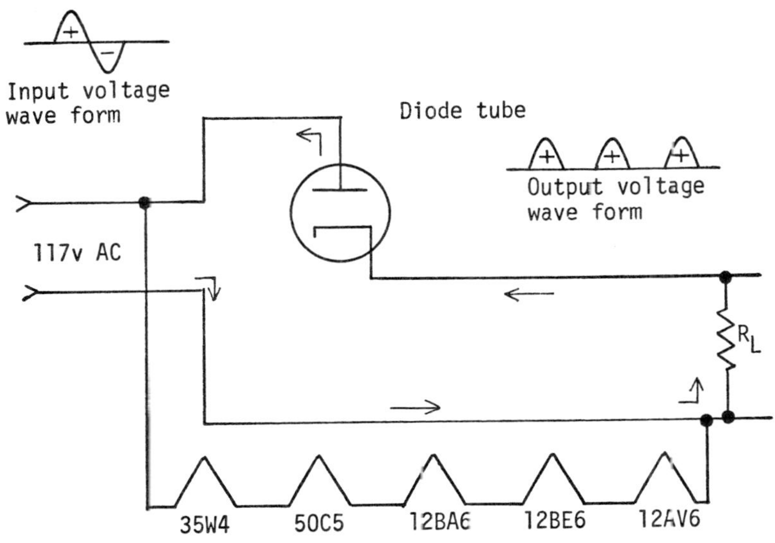

FIGURE 5-3: Basic transformerless half-wave rectifier. Shown are five tubes connected in series to the line.

line and the rectifier tube. It is used in almost all of the tube-type table model radios, and in many of the portable and table model television receivers. Figure 5-2 shows the power line connected directly to the tube's plate, and the other side of the line is connected to the cathode through the load resistor. The input wave form is shown, as well as the output wave form. The operation of this circuit is the same as the first half-wave circuit in Figure 5-1. The bottom part of the circuit shows how the filaments of the five tubes that are normally used are connected in series. The first two numbers of the tube represent the voltage of the filaments for that tube. When they are connected in series, their total voltage on the filaments equals that of the line voltage; this eliminates the use of a filament transformer.

Rectification: Full-Wave Bridge Circuit

Figure 5-4 shows a different type of full-wave circuit. One of the advantages of this type of circuit is the use of a transformer without a

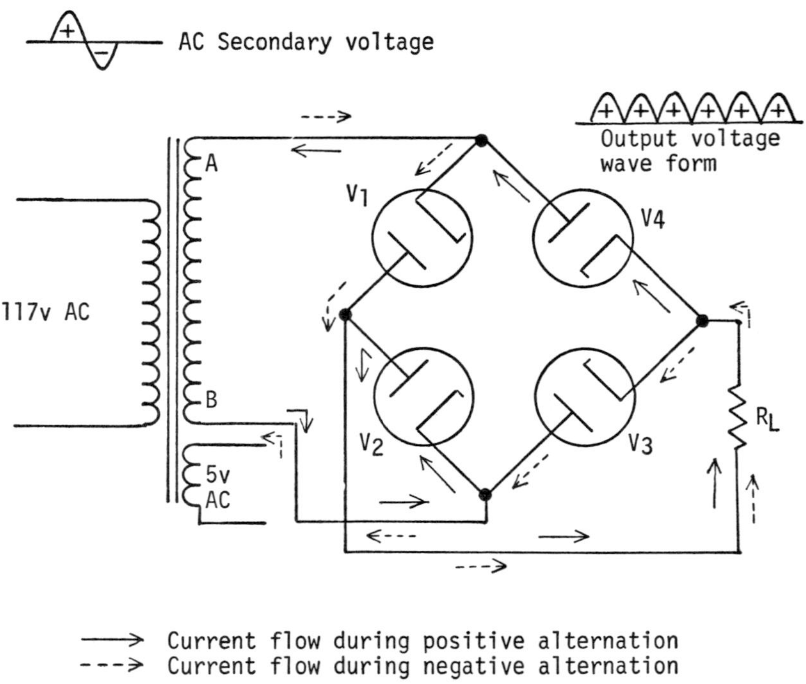

FIGURE 5-4: Bridge circuit for full-wave rectification.

center tap on the secondary winding. Another advantage is that the output voltage is twice that of the circuit using a duo-diode tube. However, the current rating is half that of the duo-diode circuit. There are four tubes used in a bridge configuration; two tubes operate during the first alternation, and the other two operate during the second alternation. During the positive alternation, the point A at the top of the transformer is positive, and the bottom end at point B is negative. Follow the solid arrows for the current flow during this alternation; current flows from point B to the intersection of V_2 and V_3. The voltage is negative, which makes the cathode negative and the plate positive, so that current will pass through the tube. V_1 has a positive voltage applied to its cathode; this makes the plate negative, so that current will not flow in that direction. The current then passes on to the resistor and up through to the intersection of V_3 and V_4. V_3's plate is connected to a negative voltage so that current will not flow through, but V_4 has a positive voltage connected to its plate so that current will pass through on to the transformer and down through to point B, from where it started. While current passed through the load resistor, it caused a voltage drop and produced an output, shown by the first positive alternation.

The AC voltage now reverses itself; this makes point A negative and point B positive. (The dotted lines show the current flow during this alternation.) Current flow now is from point A to the intersection of V_1 and V_4. V_4 has a negative voltage connected to it so that current will not flow in that direction. However, V_1 has a negative voltage connected to the cathode. This makes the plate positive and current will flow through the tube to the resistor. Current flows up through the resistor to the intersection of V_4 and V_3; V_4 has a negative voltage connected to its plate so that current will not pass through. V_3 has a positive voltage connected to its plate so that current will pass through to the bottom of the transformer and back to the top, where it completes another alternation. As current passes up through the resistor, it causes a voltage drop and produces an output as shown by the second positive alternation. Notice that the output voltage wave form is all positive alternations as they were in the duo-diode full-wave; this is also due to the undirectional current flow through the tubes.

FILTERS AND REGULATORS IN THE POWER SUPPLY

So far we have discussed and shown only the basic rectification of an AC voltage with the output wave from a pulsating DC voltage. This voltage cannot be used to operate electronic equipment; a steady non-fluctuating voltage is required, or a voltage that flows in one direction only. In order to obtain this pure DC voltage, we must use some form of filtering that will clip the peaks off the pulses and leave a straight line traveling in one direction.

The filter network is usually comprised of at least one filter capacitor; most have two, and a filter choke. The capacitance of the filter capacitors is usually high, and the inductance of the choke is also high. Whichever of these components is placed next to the rectifier, the supply is referred to that component as being the input. For example, refer to Figure 5-5(A). The choke is connected directly to the cathode; this is referred to as choke input. This type of filter network is called L type, because the components are arranged to resemble an inverted L. Refer now to Figure 5-5(B). The filter capacitor is connected directly to the cathode of the tube; this would be referred to as capacitor input.

The capacitor input is called the π network because it resembles the Greek word *pi*. The design of most power supplies includes a bleeder resistor across the output of the supply, primarily as a safety precaution for anyone touching the supply after the power has been removed. The capacitors in the supply have the ability to store voltage and will hold this voltage for many hours after the power has been removed. Placing a bleeder across the output bleeds off this voltage charge after the power has been removed. The second reason for plac-

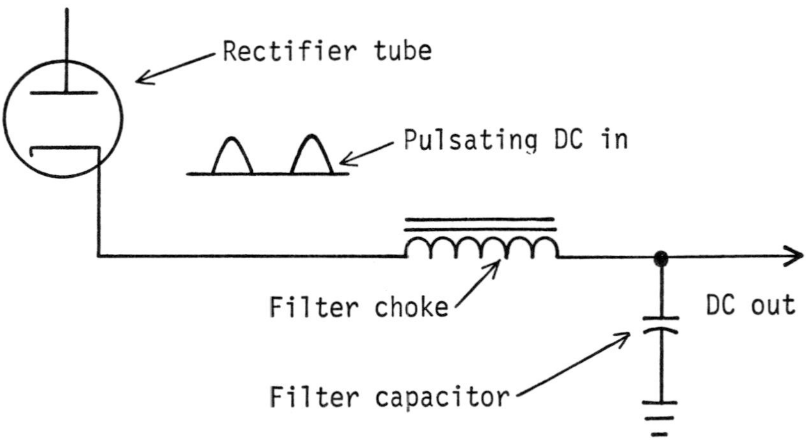

FIGURE 5-5(A): Choke input filter and L type network.

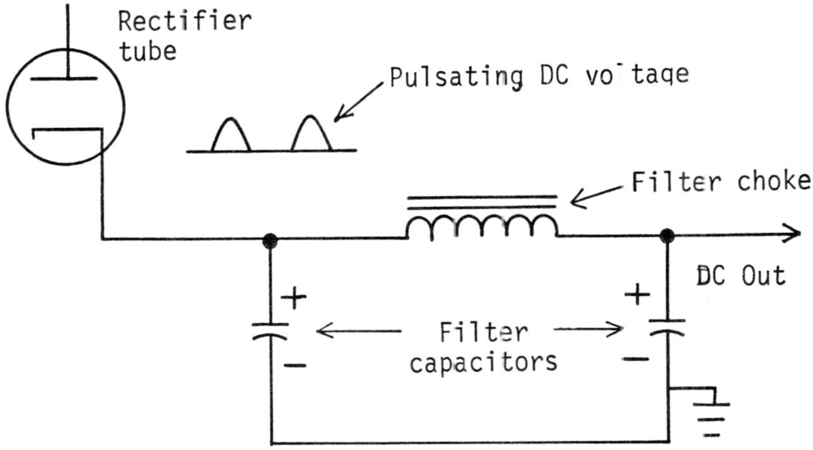

FIGURE 5-5(B): Capacitor input filter and pi network.

ing the bleeder across the output is that it places a constant load across the output with or without the regular load present. It is drawing current at all times so that when the load is placed on the output, the variation of the voltage is less. This current across the bleeder is controlling a part of the voltage so that the bleeder is acting like a voltage regulator. Regulation in a power supply is the ratio between the no-load and the full-load, and is given as a percentage.

The formula for finding this percentage of voltage regulation is as follows:

$$\text{Percent} = \frac{(En1 - En)\,100}{En}$$

En^1 = no-load voltage
En = full-load voltage
% = percentage of regulation

Example: A power supply with a DC rating of 450 volts. When the electronic equipment is connected to it, the voltage drops to 400 volts. What is the percentage of regulation?

$$\% = \frac{(450 - 400)\,100}{400} = \frac{50\,(100)}{400} = 12.5\%$$

In the tube type power supply, better voltage regulation can be maintained by the use of a gas-filled tube called the VR tube, or voltage regulator tube. Some of the more popular numbers of these tubes are OA3/VR-75, OB3/VR-90, OC3/VR-105, and OD3/VR-150.

The last numbers following the letters VR indicate the amount of voltage this tube will regulate. For example, VR-75 will regulate a constant voltage at 75 volts. Ionization takes place about 30% higher than the tube rating. When applied voltage exceeds this amount, the tube breaks down (ionization takes place) and the plate voltage is held at 75 volts. The maximum amount of current this tube is capable of is approximately 40 ma. When the plate current falls below 5 ma., ionization ceases and the tube will not function. Solid state power supplies and regulators will be covered in a later chapter.

USING THE OSCILLOSCOPE TO TEST POWER SUPPLIES

We have described rectification of an AC sine wave to a pulsating DC wave form. We have described the filter network which transforms the pulsating DC to pure DC, but suppose we could see this action taking place on the oscilloscope screen? This would give us a better understanding of this phenomenon and also make us more aware of the usefulness of the oscilloscope.

Refer to Figure 5-6. This is a half-wave power supply, using a power transformer, a half-wave rectifier tube and a π filter network. The voltage on the secondary of the transformer does not have to be exactly that shown, but make sure that this voltage does not exceed the rating of the capacitors. Four points on the supply are shown—points A, B, C, and D—and above the diagram is shown four wave forms that you should observe during this test.

First connect the positive side of the oscilloscope to point A and ground as shown. Apply power to the supply and set the oscilloscope sweep to 60 Hz. You may have to reduce the vertical gain control on the oscilloscope. Adjust the oscilloscope controls for one or two Hz; if you have a single sine wave displayed, you are now observing one Hz of the 60 Hz across the secondary of the transformer. This is shown at first wave form A. Now remove your probe and place it on the primary or input of the transformer (this is not shown). Notice you have the same sine wave but the amplitude is reduced. This then proves that the transformer is a step-up, because the amplitude of the voltage is greater in the secondary than it is in the primary. Remove the power from the supply and break the circuit at point B. This removes the complete filter network from the supply. Insert a 10k, 10 watt resistor between

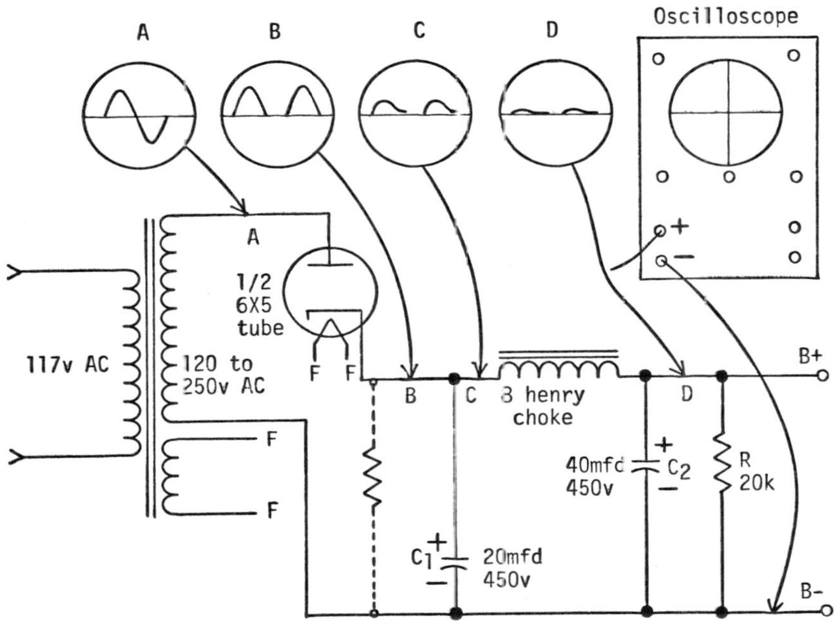

FIGURE 5-6: Half-wave power supply showing scope connections and wave forms to test effectiveness of filter network.

the cathode and B−, and then connect the positive oscilloscope probe to the cathode and the resistor. Apply power to the supply. Notice that you have to increase the vertical gain on the oscilloscope. This shows a loss of voltage. This is the loss across the rectifier tube and the voltage drop across the resistor.

Notice that we have only positive alternations, with a space between them; this is half-wave rectification, and the wave form should look like that at B, or the second wave form shown. The positive alternations you are viewing were produced when the plate of the rectifier was positive and the cathode was negative. When the AC reversed, the plate was negative and the cathode was positive. The tube would not conduct and voltage was not produced in the output. This is the blank space you see between the positive alternations.

Remove the power from the supply and the resistor from cathode to B−. Replace the capacitor C_1 to the cathode; disconnect the choke at point C. Now we have only one capacitor in the filter network. Connect the positive oscilloscope lead to point C, apply power to the supply, and the wave form should look like that at C, above the circuit. Notice that the amplitude of the wave form has been reduced. We might say it has been clipped off. This voltage was used by the capacitor to charge up to the source voltage; this is the action of the filter capacitor. Remove the power from the supply. CAUTION: The filter capacitor has the full charge of the supply on it. Use a screwdriver blade between the capacitor and B−. This will discharge the voltage. You will see a medium-size spark discharge from the blade when you do this. This is the voltage discharging.

Connect the choke back to point C; now we have the complete filter network connected back into the circuit. Connect the positive oscilloscope lead to point D and apply power to the supply. You may see nothing but a straight line. This would indicate an almost pure DC voltage. Turn the vertical gain control up on the oscilloscope and you will probably see near that in D, at the top of the circuit. The fact that you had to turn the gain of the oscilloscope up to see any ripple in the DC voltage shows that the filter network performed as it should. Remove power from the supply, disconnect C_2 from the circuit, and leave the oscilloscope lead at point D. Return the vertical gain control on the oscilloscope to where it was when you first viewed point D. Apply power to the supply, and the wave form should look very near that between C and D. This test simulates the situation you might find if, in checking some form of electronic equipment with your oscilloscope, you noticed this type of wave form, indicating your trouble might be in the power supply. If this capacitor were to open up, this would give you some idea of where to look. You would then disconnect the supply from the equipment, place the oscilloscope as you have it now, and you would see this type of pattern. If C_1 were to open up instead of C_2, the pattern would be near that of C wave form for half-wave rectification.

Figure 5-7 shows a full-wave power supply with oscilloscope connected and wave forms for different points on the supply. This test for this supply is exactly as for the half-wave supply just explained. The main difference between the two is that in the full-wave supply, both alternations are rectified. In wave form B, notice there is no spacing between the alternations. Notice too, that the inductance of the choke is smaller and the values of the capacitors are also smaller. The frequency of the full-wave supply output is 120 Hz; therefore, it does not require as much filtering as the half-wave supply, which had an output frequency of 60 Hz.

The same troubleshooting technique is used with this supply as was used in the half-wave explanation.

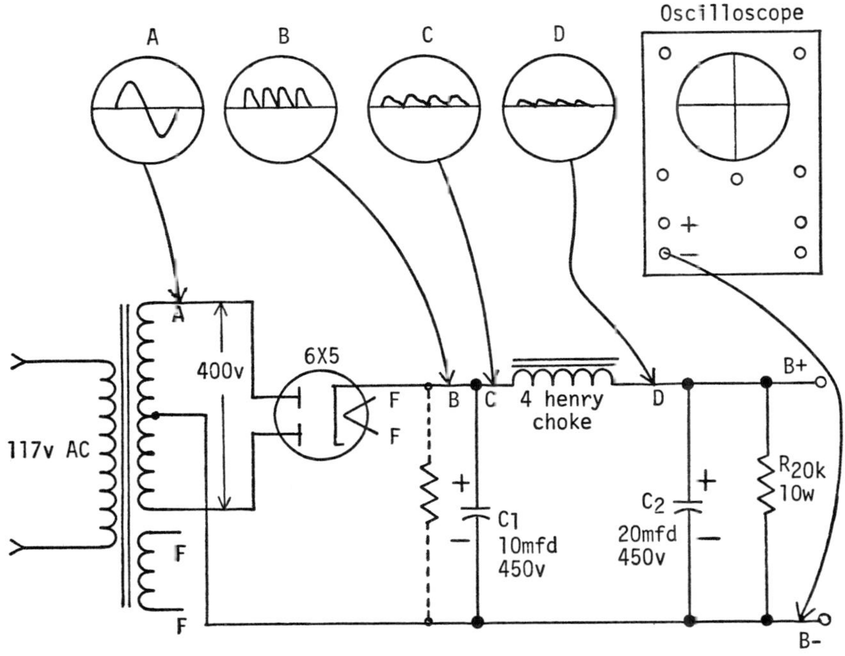

FIGURE 5-7: Full-wave power supply showing scope connections and wave forms to test effectiveness of filter network.

6

Communication—Sending: Amplitude Modulation

AMPLITUDE MODULATION DEFINED

Amplitude modulation is defined thus: modulation in which the amplitude of the carrier-frequency current is varied above and below its normal value in accordance with the audio, picture, or other intelligence signal to be transmitted.

Modulation refers to the term *modulate*, which means to vary the amplitude. The term *carrier-frequency* refers to the RF signal transmitted by a transmitter and unmodulated. The carrier is generated by the oscillator circuit in the transmitter and then usually amplified. The term *intelligence* refers to the modulation applied to the RF carrier, such as music, voice, or picture.

Let us examine this process more closely. You recall in Chapter 4, in our study of oscillator circuits, we used the oscilloscope to examine the RF voltage produced by an oscillator circuit. The pattern produced on the oscilloscope was a steady amplitude of sine waves as long as we did not change the frequency. When we did change the frequency, the amplitude did not change; only the spacing between the sine waves changed. This then is what we would refer to as the carrier-frequency or unmodulated RF carrier.

Turn to (a) in Figure 6-1. This illustration shows the RF carrier. Notice that the negative and positive alternations have the same amplitude as well as the same space between the wave forms. (b) in Figure 6-1 shows an audio frequency wave form. This is a continuous frequency of around 400 Hz. If the frequency were varied like that of music or voice, each wave form would have a separate amplitude and the spacing between the wave forms would be different. In a radio wave this frequency would be called the intelligence, because it is the information that is produced at the studio and then reproduced in your home in the form of music, voice or TV pictures.

If we combine this RF wave form and AF together, we will have a radio or TV signal that can be transmitted. Figure 6-1 shows in (c) what the radio signal looks like when it is transmitted; the difference in amplitude is caused by the variation of the AF frequency. This variation in amplitude from maximum to minimum forms what is referred to as the modulation envelope.

All communication is in the form of an AC wave form, and when making reference to the wave form, we use a zero reference which is a

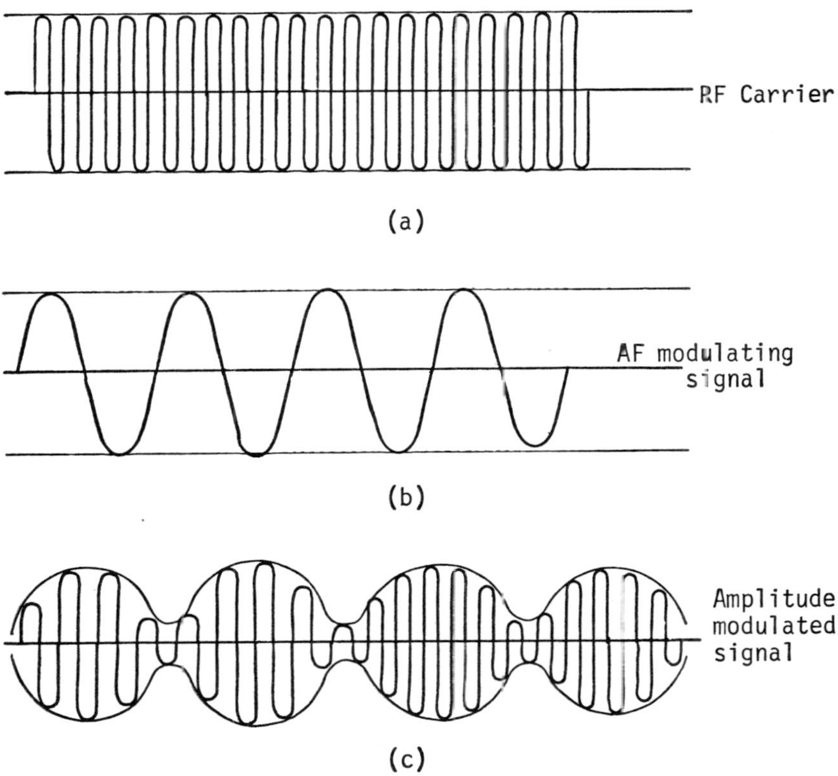

FIGURE 6-1: RF carrier, AF modulating signal, and amplitude modulated signal.

straight line. Everything above this line is positive and everything below it is negative. In referring to the modulated envelope, everything above this line we call the upper side-band because it is the upper side, or above the line, which contains the RF and AF. Everything below this line is called the lower side-band; it also contains the RF and AF. Now we can say that everything above the reference line is the upper side-band, and everything below the line is the lower side-band.

The RF carrier does not change in frequency, but the modulating signal (audio) does change in frequency and in amplitude. This changing audio is superimposed on the RF carrier and this forms the modulating envelope, which consists of the upper and lower side-bands. Therefore, as stated earlier in this chapter, the intelligence is in the side-bands.

If a single tone of audio or if a single frequency of audio is used to modulate a carrier, there are two other frequencies produced. For example: An RF carrier of 1400 KHz is modulated by a 1000 Hz audio frequency; now the upper side-band will be the carrier plus the audio. 1400 KHz + 1000 Hz = 1401 KHz; this is the new frequency of the upper side-band. The lower side-band will be the carrier minus the audio frequency or 1400 KHz − 1000 Hz = 1399 KHz.

Figure 6-2 illustrates in (a) the 1400 KHz RF carrier which is not modulated. Notice that the amplitude and the spacing between the wave forms are all the same. (b) shows the upper side-band, which is the RF carrier plus the audio frequency. Notice that the wave forms are close together, which means that we are going higher in frequency from 1400 KHz to 1401 KHz. (c) shows the lower side-band which is the carrier minus the audio frequency or 1400 KHz − 1000 Hz which is 1399 KHz. Notice that the wave forms are wider apart, indicating that we are going lower in frequency. If we put these three wave forms together we would have a modulated envelope similar to that shown in Figure 6-3. The amplitude of the unmodulated carrier is marked by the arrows, above and below; this amplitude is the upper and lower side-bands. This complete wave form is the modulated envelope.

An amplitude modulated signal is produced by the process of heterodyning, or mixing the carrier frequency and the audio frequency. This process produces the upper and lower side bands. In the example we have just explained, we had a 1400 KHz carrier; this was modulated by a 1000 KHz audio signal. The process of mixing these two signals produced an upper side-band with a frequency of 1401 KHz and a lower side-band with a frequency of 1399 KHz.

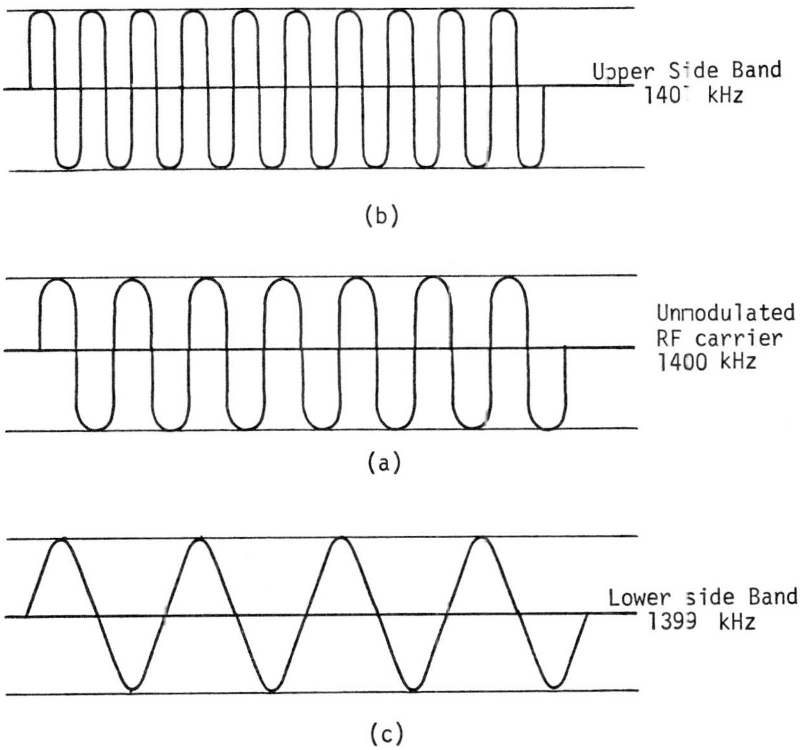

FIGURE 6-2: Unmodulated and modulated wave forms of RF carrier and upper and lower side bands.

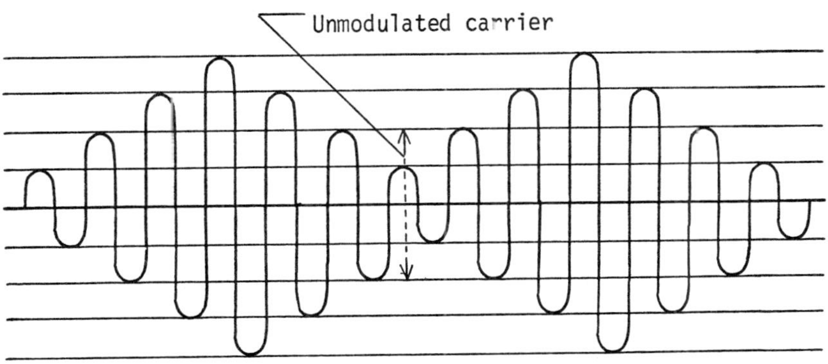

FIGURE 6-3: Amplitude modulated signal with varying amplitude.

The band width of an amplitude modulated signal is determined by the difference in frequency between the lowest frequency side-band and the highest frequency side-band signal. We could also say the band width is equal to twice the highest modulating frequency. In our example, the lowest frequency was 1399 KHz and the highest frequency was 1401 KHz. The band width would be 1401 − 1399 = 2 KHz, or highest frequency × two. This would be 1000 KHz × 2 = 2 KHz.

THE AMPLITUDE MODULATION TRANSMITTER

You have learned in this chapter how the RF carrier was modulated and how it dealt with an AM signal. (We used the abbreviation AM when referring to the amplitude modulated signal.) There are many different types of transmitters; in this chapter we will study only the type in which the intelligence is either voice or music.

In any type of transmitter, the frequency of the carrier is determined by the oscillator circuit. In commercial transmitters the oscillator frequency is determined by a crystal. To insure the stability of the frequency produced by the crystal, it is placed in an inclosure that is called an oven. A small heating element is placed inside this oven, which is attached to a thermostat, and power is applied all the time. The temperature inside the oven is controlled when it drops below a certain temperature. This keeps the temperature on the crystal at the correct operating point; whenever the circuit is energized, the crystal will operate. Since the oscillator does not produce a very large output, it is usually followed by a buffer stage. This is an RF amplifier that performs two functions.

1. It amplifies the small signal from the oscillator stage.
2. It isolates the oscillator stage from the final stage.

Since the last RF stage is usually operated at very high gain, if the oscillator were connected direct to this high gain stage, it would cause the oscillator frequency to shift with modulation. The buffer stage is usually followed by the final stage, where the power of the carrier is determined.

The audio stages usually constitute a speech amplifier. The microphone and record player are fed into a preamplifier and then into a mixer, and this is fed into the speech amplifier. The output of the

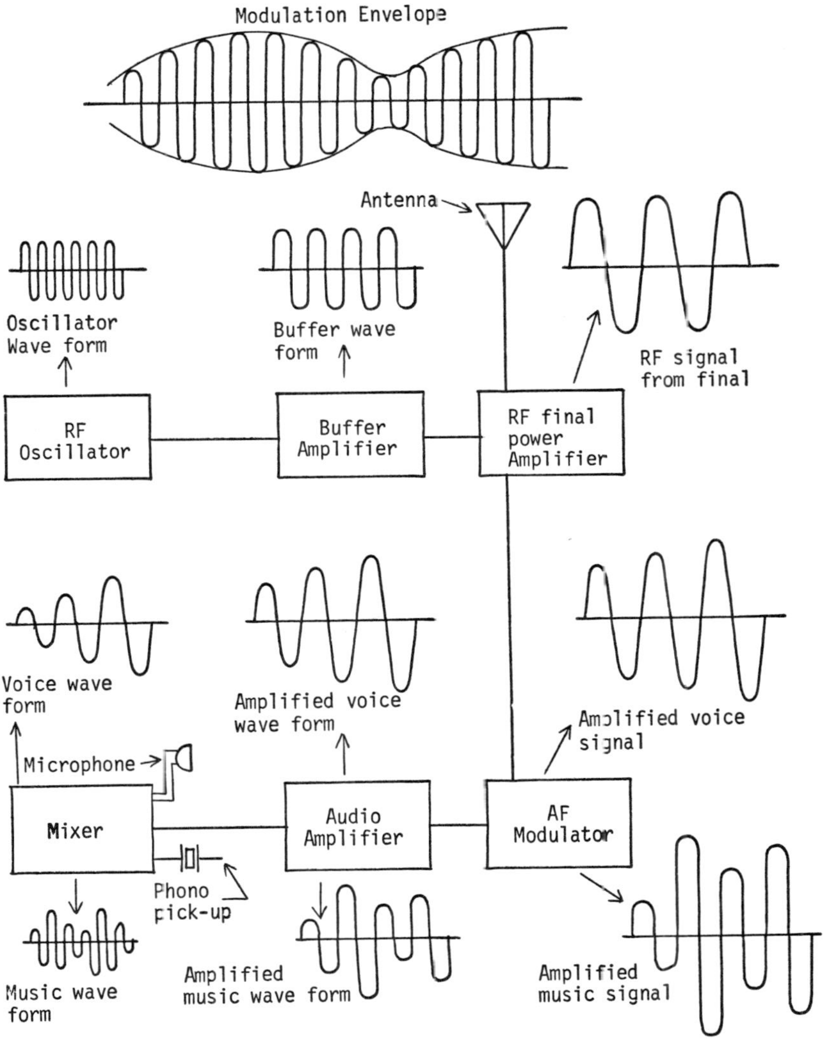

FIGURE 6-4 Basic amplitude modulated transmitter with wave forms for each stage, and the radio wave at the antenna.

speech amplifier is fed into a modulator stage; this stage is a power amplifier. Here the audio is amplified to a necessary amplitude to modulate the carrier to 100% modulation. A block diagram, shown in Figure 6-4, includes the wave form above each stage. This shows what each stage is producing and how they are combined in the final stage to produce a modulated carrier. This modulated carrier is called a radio wave, as it is transmitted out into the atmosphere on its way to the radio receiver.

This transmitter would be operated in the medium and short-wave bands, up to a power of 1000 watts. Where more power is needed, there would have to be more stages in the RF section and also in the audio or modulator stages.

In the basic transmitter that was just described, all of the stages were operated on the fundamental frequency. That is, the frequency of the crystal oscillator circuit, the buffer stage, and the RF final stage were all tuned to this frequency. This frequency might be in the broadcast band. Now, suppose we wanted to operate an AM transmitter on a higher frequency. The stability of the frequency of the oscillator stage is greater in the lower frequencies than in the higher frequencies. It is the practice, when operating a transmitter in the higher frequencies, to operate the oscillator stage at a lower frequency and then use doubler or tripler stages.

Suppose we wanted to operate an AM transmitter in the 144 MHz band. Refer to Figure 6-5. This illustrates how the oscillator is operated down at 8025 KHz. The next stage is tuned to three times the oscillator frequency (tripler stage). This would give us a frequency of 24,075 KHz. The next stage would be tuned to three times this frequency (tripler stage), which would give us a frequency of 72,225 KHz. The final stage would be tuned to twice that frequency (doubler stage); now the output stage is operating at 144,450 KHz, or 144 MHz. We could follow this stage with an RF power amplifier and tune it to the same frequency, giving a greater power output. When operating a

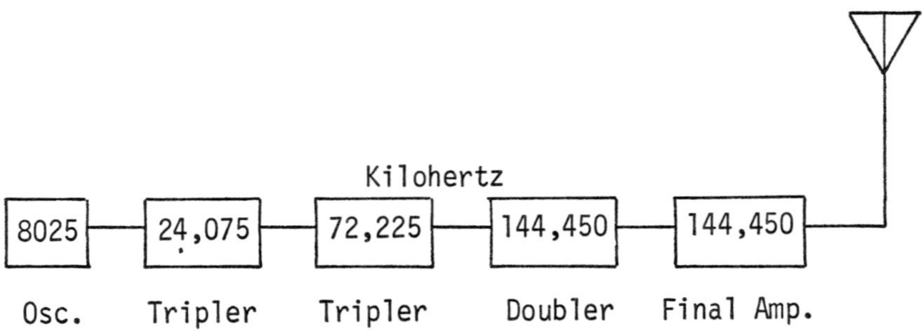

FIGURE 6-5: A block diagram illustrating a transmitter with the oscillator on 8025 KHz and the final on 144,450 KHz.

transmitter in this manner, extreme care should be taken in design and tuning. One of the harmonics generated by one of the tripler or doubler stages might be strong enough to be transmitted along with the fundamental frequency. Suppose you were a radio amateur with a license to operate in the 144 MHz band. The first tripler stage, with its frequency of 24,075, does not fall in any of the assigned amateur bands, but might be transmitted along with the 144 MHz frequency. This would be a violation of your license—and besides, you might be causing interference to another station that was authorized to operate on this frequency. There are methods by which harmonic radiation can be suppressed, but space does not permit explaining them here.

TRANSMISSION OF THE RADIO WAVE

The term transmission is defined as a message, signal, or other form of intelligence that is being transmitted. In the transmission of an amplitude modulated transmitter, the intelligence would be music or voice, for entertainment purposes, or it might be a continuous tone for test purposes. When we refer to transmission, sometimes called the signal of an amplitude modulated transmitter, we are referring to the radio wave. The transmission from the transmitter is carried to an antenna and radiated into space as a electromagnetic wave. You may possibly have observed a piece of metal that had been heated until it was so hot that you could see the heat coming off. This, of course, is heat waves, but the electromagnetic waves being transmitted from the antenna resemble these heat waves that are emitted from the hot metal. The difference is that you cannot see the radio waves as you can heat waves. These radio waves travel outward from the antenna into space; some travel close to the ground; others travel skyward. The ones that travel toward the sky are reflected back to earth by the different layers of atmosphere and are picked up by a receiving station along with some of the ground waves. This basically is the path of the transmission from the transmitter to the receiver.

TROUBLESHOOTING AM TRANSMITTERS

We mention this phase in AM transmitters only to let you compare the troubles at the transmitter and to observe that they are not much different from those in the receiver.

Tube failure is probably one of the most common troubles, just as it is in the receiver. There are transformers, chokes, resistors and capacitors in the transmitter, the same as in the receiver. The ratings may be higher, but the method of finding the trouble is almost the same. There is one advantage in the transmitter over the receiver for finding the trouble—almost every stage has a current meter and some have a voltmeter. This makes locating the trouble simpler in the transmitter. The meters on the front panel will help in spotting the stage in which the trouble is located. In the radio receiver, the speaker will tell you if there is sound or not. In the transmitter output, the antenna current meter will tell you if there is power going into the antenna. This is the main thing to remember about any troubleshooting in electronic equipment: apply what you have learned in a logical manner.

7

Communications—Receiving: AM Receivers

THE TUNED RADIO FREQUENCY (TRF) RECEIVER

In the previous chapter we learned how the radio wave was formed and how it was transmitted from the antenna out into space. This radio wave is called the modulated RF signal. At the receiving end, the antenna intercepts the signal. Recall that the signal is an electromagnetic field or wave, which is induced into the receiving antenna similarly to the way an EMF is induced into the secondary winding of a transformer by the primary. This signal produces a voltage into the antenna which is a conductor. This signal voltage is very weak (it is measured in microvolts), so it must be amplified by an RF amplifier.

The receiver performs the following functions:

1. It amplifies the weak RF signal.
2. It detects (separates) the RF from the intelligence.
3. It amplifies this intelligence so that it can be used.

The TRF receiver or tuned-radio-frequency has three basic sections. The RF amplifier is the first stage where the weak signal picked up by the antenna is amplified. A tuned circuit consisting of a variable capacitor and an inductor selects the frequency or station and applies it to the RF amplifier. The voltage from the RF amplifier is inductively coupled by transformer action to another tuned circuit, which is tuned to the same frequency as the input to the RF amplifier. The secondary of this transformer is coupled to the detector stage, which passes the

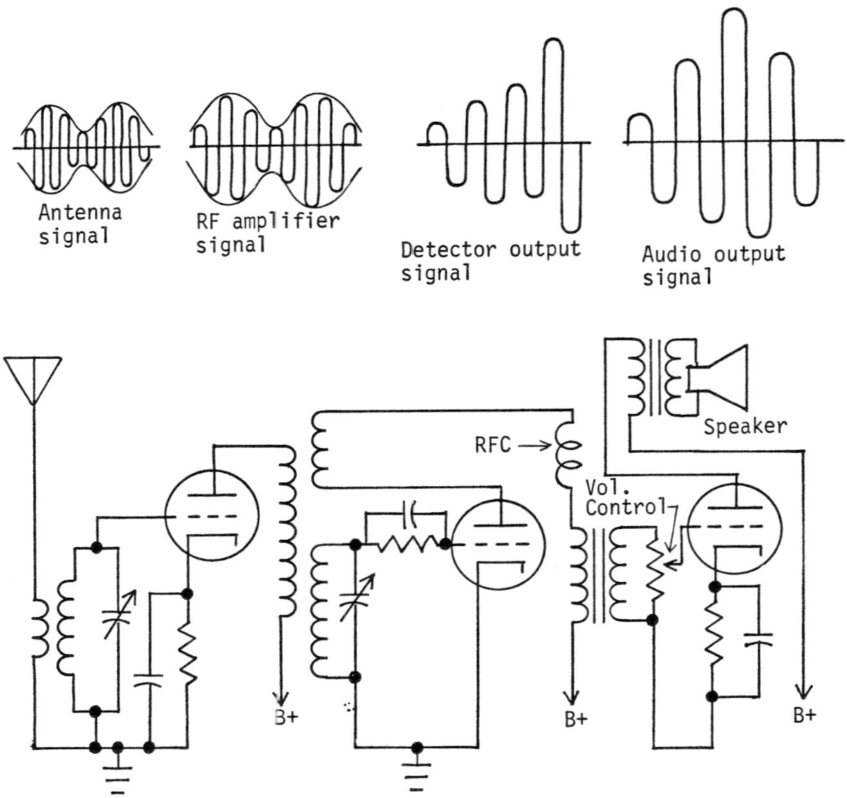

FIGURE 7-1: Basic TRF receiver with wave forms showing the output of each stage.

audio frequency and rejects the RF frequency. This audio frequency voltage is amplified and coupled to the primary of an interstage transformer and induces a voltage into the secondary winding. The secondary is coupled to the audio output stage where it is amplified and coupled to the speaker through an audio output transformer. Here it is produced as sound. Figure 7-1 illustrates a basic TRF receiver with the wave form as it appears in the antenna through each stage to the speaker.

SUPERHETERODYNE RECEIVER

A superhetrodyne receiver is defined as a receiver in which all incoming modulated RF carrier signals are converted to a common IF carrier value for additional amplification and selectivity prior to demodulation, using heterodyne action. The output of the IF amplifier is then demodulated in the second detector to give the desired AF signal. The definition of the term by which this action is accomplished is "*heterodyne*: designating the combination of two alternating current signals of different frequencies corresponding respectively to the sum of and the difference between the two original frequencies." This definition will apply to AM, FM, and TV receivers. They all use the heterodyne action. I will describe the latter two in later chapters. In this chapter I will discuss the AM receiver, using a block diagram to illustrate the main stages in the superheterodyne receiver.

Some receivers use an RF stage and some do not, but we will say the better receivers use the RF stage, which increases the selectivity and sensitivity of the receiver. Figure 7-2 illustrates the block diagram of the superheterodyne receiver using the RF stage. The wave form from the antenna through each stage is shown. Notice the wave form of the oscillator and that each cycle has the same amplitude and spacing between cycles.

HETERODYNE OPERATION

Before we break down the definition of the term *heterodyning*, there are a few things we should know first.

A receiver with an RF stage will have a three-ganged tuning capacitor. The RF stage, the mixer-converter, and the oscillator stages all will have separate tuning or variable capacitors, which will be connected by a common shaft. When the shaft is rotated, all of the capacitors will turn at the same time. The second point is that all AM receivers use a standard intermediate frequency (IF) that is 456 KHz (a few transistor car radios are still using the old 262 KHz IF). One last point is that almost all modern day AM superheterodyne receivers (tube type) use a pentagrid-converter tube, which is a special purpose type tube. In the block diagram the oscillator is shown separately, but these are the components of the oscillator circuit.

Now back to the block diagram 7.2. We turn the receiver on, and immediately the oscillator starts to oscillate. The frequency of oscillation will depend upon where the tuning capacitor is setting. Start rotating the dial; at 1400 KHz the antenna has intercepted a signal, which is made resonant by the variable capacitor and coil in the grid circuit of the RF stage. The signal voltage is fed to the grid and

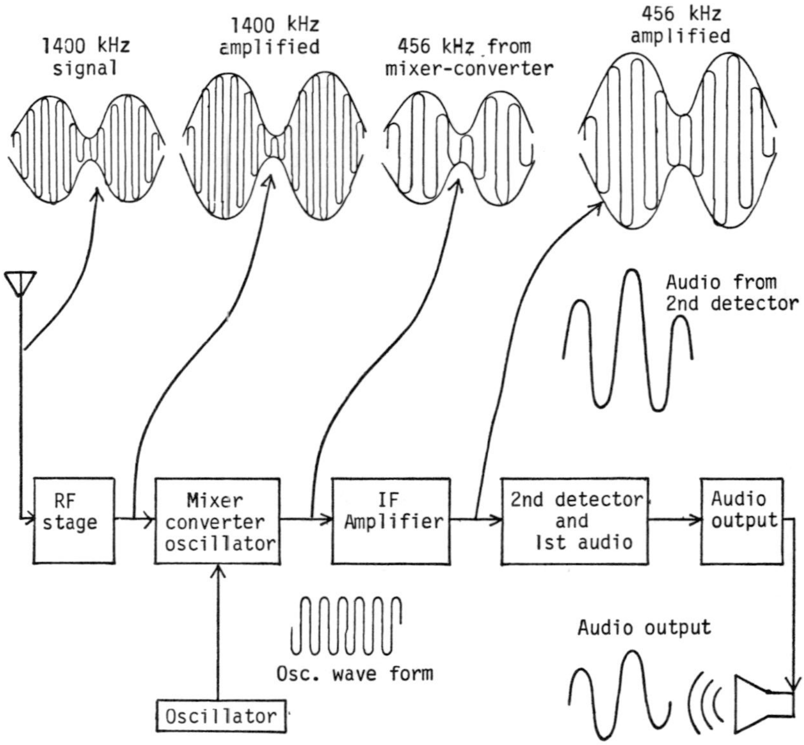

FIGURE 7-2: Block diagram of a superheterodyne receiver, and the wave form for each stage.

amplified in the tube. It is then coupled to the grid of the mixer-converter stage by transformer action from the plate of the RF amplifier. The secondary of this transformer is tuned by the variable capacitor across this coil. Now the grid of the mixer-converter stage is resonant at 1400 KHz signal. Since the variable capacitor in the oscillator stage was rotated at the same time, the variable in the mixer stage was rotated. The oscillator frequency now is 1856 KHz. The oscillator voltage is fed to the oscillator grid of the tube at the same time the RF voltage is fed to the RF grid. Now the two voltages are mixed or one voltage beats against the other, and a new frequency is formed at the plate of the tube. This new frequency is the difference between the two frequencies, which is 456 KHz. 1856 − 1400 KHz = 456 KHz: this frequency is fed to the IF transformer primary, which is resonant at this frequency.

A point to remember here is that the oscillator frequency during heterodyning action is always the IF frequency above the Rf frequency.

The definition of the term heterodyning is as follows: producing the third frequency through the sum or difference of two frequencies.

From the plate of the mixer-converter the signal is fed to the primary of the IF transformer, and by mutual induction, the signal voltage is induced into the secondary winding of the transformer. The secondary is connected to the grid of the IF amplifier. Here the 456 KHz signal is amplified and the plate is connected to the primary of the second IF transformer. Here again, by mutual induction, the signal voltage is inductively coupled to the secondary of the transformer. Each IF transformer is pretuned at the factory to the 456 KHz frequency, but each channel on the broadcast band has a band width of 10 KHz. Both transformers and the IF amplifier must pass a band width of 10 KHz. If the center frequency of the IF is 456 KHz, the upper side-band will swing to 461 KHz. The lower side-band will swing to 451 KHz. This gives us a band width of 10 KHz, or $461 - 451 = 10$ KHz. All frequencies above and below this 10 KHz band width are attenuated.

The secondary of the second IF transformer is coupled to the second detector stage; this stage removes the intelligence from the carrier. The second detector tube is usually a special-purpose tube called a duo-diode-triode. It has two diode plates that are common to one cathode, and a triode section which acts as the first audio amplifier. The secondary of the second transformer is coupled to the diode plate; this section acts as a detector which has a resistor and capacitor network connected to its output. This network acts as a filter, passing the low frequencies and rejecting the high ones. The low frequencies are the audio; the high frequencies are the RF. The output of this network, which is the audio, is fed to the volume control. The volume control is coupled to the grid of the triode section of the duo-diode-triode tube. The signal is amplified and capacitive coupled to the output audio amplifier. The secondary of the second IF transformer is capacitive coupled to the second diode of the second detector tube. It is rectified, and the output is coupled to a filter network, which filters the pulsating DC voltage. This DC voltage is fed back to the grid of the first RF amplifier, the mixer-converter, and the first IF amplifier. This is the AVC voltage, which was developed from the RF carrier and is used as a bias on the first three stages. The letters AVC refer to automatic volume control; you can see now that it does automatically control the volume. The carrier is an AC voltage, picked up by the antenna, amplified through the first three stages, then rectified and filtered, and used as a bias to control the amplification of the first three stages, which control the audio amplification or volume. This is done automatically. The stronger the incoming signal, the more bias is developed, thus increasing the bias on the first three stages. This will decrease the amplification, which will decrease the volume of the audio stages.

RECEIVER STAGE ANALYSIS

Radio Frequency Stage

You have followed the signal from the antenna through the receiver to the audio output and the speaker. I will go back now and analyze each stage, explaining the function of the parts used.

Figure 7-3 illustrates an RF amplifier or the first stage in the receiver. The antenna is connected to the antenna coil (T_1) primary; the RF voltage is inductively coupled to the secondary of T_1, which is connected through a resistor R_1, which is connected to the AVC bus line. C_1 is a decoupling capacitor, which couples the RF to ground, preventing it from going back into the AVC circuit. C_2 is a variable

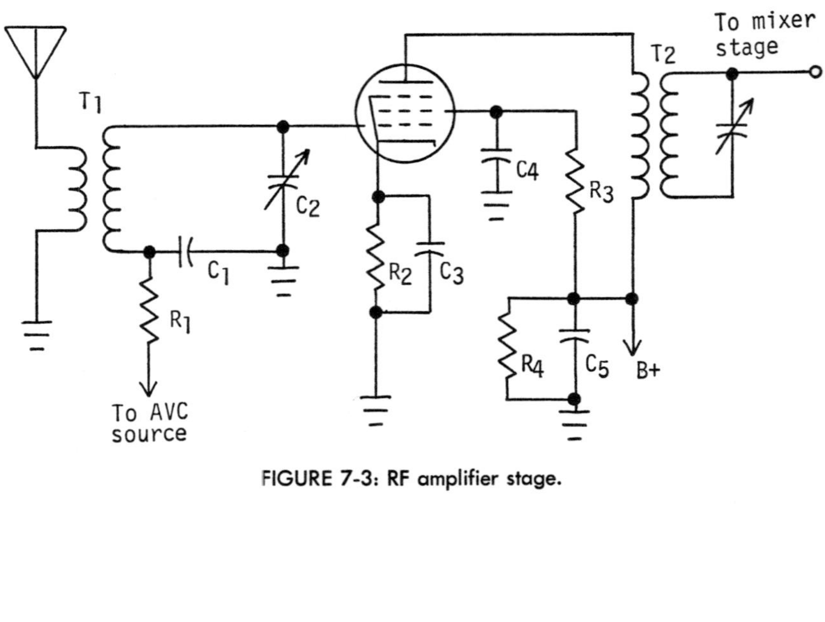

FIGURE 7-3: RF amplifier stage.

capacitor, which when combined with the secondary of the antenna transformer T_1 forms a tuned circuit that selects the incoming stations. R_2 is the cathode resistor; the voltage developed across this resistor is the grid bias for the tube. Since the cathode resistor R_2 is the grid bias source, the capacitor C_3 is the cathode bypass capacitor. Its function is to bypass the RF in the cathode circuit around the grid bias power supply. This capacitor also performs a second function: a capacitor has the ability to store a voltage over a period of time and then tries to prevent any change in voltage across it. Then it is acting as a voltage regulator for the bias supply.

C_4 is the screen decoupling capacitor which couples the RF in the screen circuit to ground, preventing it from entering the power supply. R_3 is the screen dropping resistor; the voltage drop across this resistor supplies the screen with the necessary operating voltage. R_4 is a part of a voltage divider network formed by R_4 and R_3; its purpose is to help regulate the screen voltage. C_5 is connected to the lower end of the primary of the RF transformer T_2; its purpose is to prevent any RF at this point from entering the power supply. T_2 is an untuned RF transformer which couples the amplified RF signal to the grid circuit of the mixer-converter stage.

Oscillator Converter

This stage is sometimes referred to as the detector-oscillator stage. In this stage the heterodying action takes place in the receiver. In some of the earlier receivers, the oscillator is a separate tube with its associate circuits. Due to the cost, most of the receivers use one tube called the pentagrid-converter tube.

Figure 7-4 illustrates a mixer-converter stage. T_1 is an RF transformer coupling the signal from the preceding stage to grid G_3 of the mixer-converter stage. The resistor R_1 couples the AVC voltage to grid G_3 of the tube. The capacitor C_1 is a decoupling capacitor in the AVC network to the grid. Capacitor C_2 is a variable capacitor across the secondary of T_1 transformer. This combination tunes the signal at grid G_3 of the mixer tube to the same frequency as that of the RF stage. C_{2a} is a trimmer capacitor across the variable C_2. It is used when aligning the RF section of the receiver, and for tracking or aligning the tuning section through which the signal passes so that the dial calibrations will be correct. (We will discuss alignment later in the chapter.)

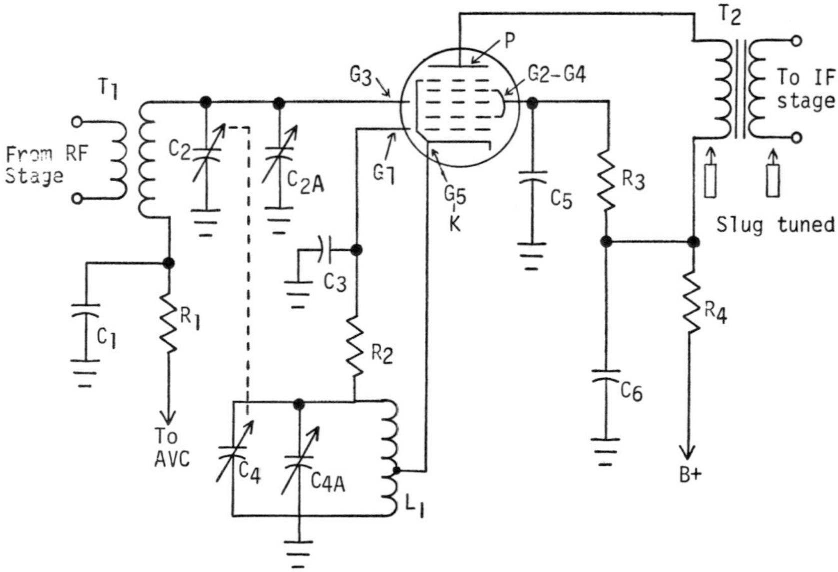

FIGURE 7-4: Mixer-converter stage.

Resistors R_2 and C_3 are the bias coupling network from the source to the grid G_1 of the tube. The bias voltage is the oscillator voltage developed by the tuned circuit L_1 and C_4, which is the oscillator frequency. The capacitor C_{4a} is a trimmer capacitor across capacitor C_4. It is used for tracking the oscillator when aligning the receiver. (This also will be covered later.) The dotted line from C_2 variable capacitor to C_4 indicates the way shafts of these two capacitors are fastened together. The point on the indicator L_1 that is coupled to the cathode of the tube is the oscillator feedback.

This is a Hartley series tuned oscillator circuit, which puts the lower part of the inductor from the point where the cathode is connected to ground in series with the power supply. The positive end of the power supply is connected to the plate which is the output of the oscillator. The feedback voltage is connected to the lower end of this inductor; this voltage is then inductively coupled to the inductor which is the plate coil. The feedback voltage in the plate coil is inductively coupled to the grid coil; this voltage is in phase with the grid voltage and adds to it to sustain oscillation.

The capacitor C_5 is the screen decoupling capacitor; it couples the AC signal present in the screen grid circuit around the supply to ground. The voltage supply source for the screen grid is the resistor R_3. The difference between the voltage drop across this resistor and the supply voltage is the screen grid voltage, so R_3 is the screen dropping resistor.

The capacitor C_6 is the plate decoupling capacitor; it decouples the AC voltage present at the lower end of the transformer, around the supply. The supply for the plate circuit is the resistor R_4. The difference between the voltage drop across this resistor and the supply is the plate voltage.

Transformer T_2 is the first IF transformer and couples the signal voltage from the converter to the IF amplifier stage.

Intermediate Frequency

Figure 7-5 illustrates an intermediate frequency (IF) amplifier. This amplifier must be capable of passing a band width of 10 KHz since this is the band width of each AM channel.

Transformer T_1 is the coupling device between the converter and the grid circuit. By mutual induction, the signal is coupled into the grid. Resistors R_1 and C_1 are connected to the AVC bus line which gives this stage an automatic volume control when the receiver is in operation.

R_2 is the cathode resistor; voltage developed across this resistor is the bias for operating the tube. Capacitor C_2 is the cathode decoupling

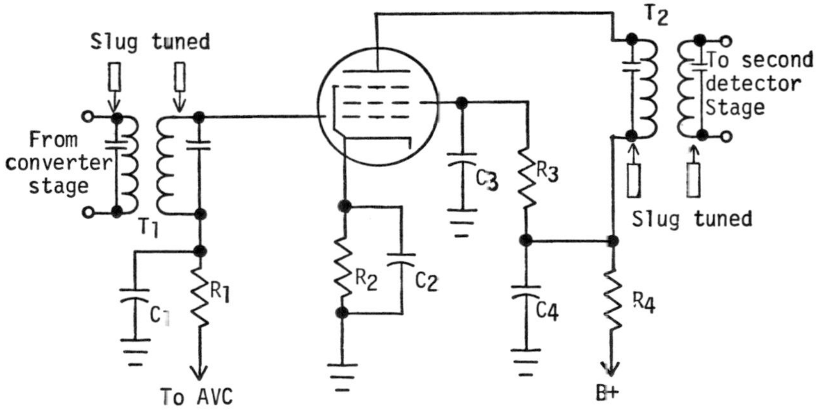

FIGURE 7-5: Intermediate frequency (IF) stage.

capacitor, which couples the signal voltage on the cathode around the supply voltage (voltage across R_2). Capacitor C_3 is the screen grid decoupling capacitor, which couples the signal voltage in the screen grid circuit around the supply voltage to ground. The supply voltage for the screen grid is developed across resistor R_3. The difference between the voltage developed across this resistor and the supply voltage is the screen voltage. Capacitor C_4 is the plate decoupling capacitor. The signal voltage that appears at the point of T_2 and R_4 is coupled to ground by this capacitor. The plate load resistor R_4 supplies the plate voltage; the difference between the drop across this resistor and the supply voltage is the plate voltage. Transformer T_2 is the IF transformer which couples the signal from the plate of the IF amplifier and the second detector.

SECOND DETECTOR, AVC, AND FIRST AUDIO STAGE

Figure 7-6 illustrates, first, the second detector stage. The trans-

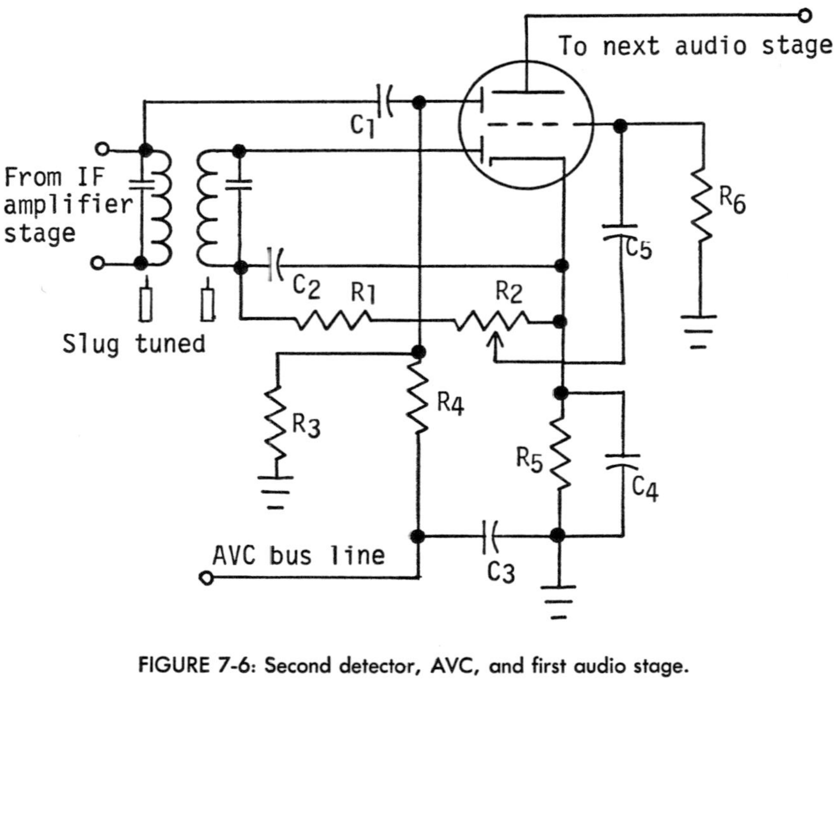

FIGURE 7-6: Second detector, AVC, and first audio stage.

former T_1 is the coupling transformer from the IF stage. The signal is fed to the primary and mutually induced into the secondary. The secondary of this transformer is coupled to one of the diode plates which, along with the cathode, acts as a detector. Capacitor C_2 is the RF filter capacitor, and resistor R_1 and R_2 form the detector load. The output of this filter network is an audio signal which is controlled by varying the resistance of R_2. The output is coupled through a coupling capacitor to the grid of the first audio tube which is the triode section of the tube.

Automatic Volume Control

The primary of T_1 is coupled through capacitor C_1 to the second diode plate of the tube; this is the RF signal voltage. This AC voltage is rectified by the diode section, and the output is coupled through the filter network of R_3, R_4, and C_3; the output is a DC voltage that is fed back to the preceding stages as an automatic bias. The amplitude of this voltage is determined by the amplitude of the RF signal from the input of the receiver.

First Audio Stage

The triode section of the tube is the first audio stage; the audio from the detector stage is coupled through C_5 to the grid. The resistor R_6 is the grid resistor. The audio is amplified in this stage and then coupled to the next stage.

Audio Output Stage

Figure 7-7 illustrates an audio amplifier. Capacitor C_1 is coupling the audio from the first audio stage to the grid; R_1 is the grid resistor; R_2 is the cathode resistor. Voltage developed across this resistor is the operating bias for the tube. C_2 is the cathode decoupling capacitor, which couples the signal in the cathode circuit around the supply (voltage across R_2). Capacitor C_3 is the screen decoupling capacitor, which couples the signal voltage in the screen circuit around the supply. The screen supply voltage is the voltage drop across R_3 subtracted from the supply voltage. The amplified signal in the plate circuit is coupled to the primary of the output transformer, and the output of this transformer is coupled to the speaker.

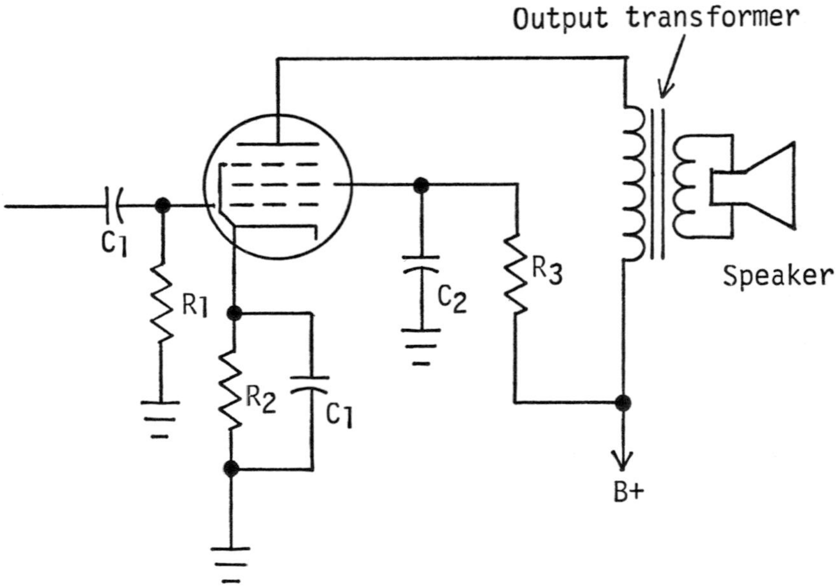

FIGURE 7-7: Audio output amplifier connected to a speaker.

TROUBLESHOOTING THE AM RECEIVER

Before removing the chassis from the cabinet, you should first determine what you are looking for. Turn the receiver on. If there is no sound coming from the speaker, look through the back, or remove the back and see if the tubes are lit up. If all the tubes are lit, you should check them for possible defects. If all the tubes check all right, the trouble is in the circuit.

Remove the chassis from the cabinet, and visually inspect the wiring and components for any visible signs. One of the resistors may show signs of overheating or perhaps it is even burned in places. Check the wires leading to the voice coil of the speaker; sometimes they will come loose from the cone.

If there are no visible signs of the defect, you must use a systematic method of finding the trouble. There are two methods to use: (1) Starting at the speaker and using an AF generator, inject a signal into each stage back to the second detector; this is the audio section. From the second detector back to the antenna is the RF section; use an RF generator and inject a signal into each stage until the defective stage is found. (2) When the defective stage is found use a VOM or VTVM and measure the resistance and voltage to find the defective component.

Figure 7-8 illustrates the audio stage of the receiver and the points where the signal should be injected, using an AF generator. The generator is shown connected to the voice coil of the speaker, to see if it is defective.

Next, it should be connected to the grid of the audio output tube, marked with an arrow and letter B. Now move the generator lead to the center tap of the volume control, which is marked with an arrow and letter C. With the generator connected to point A, if no sound is heard in the speaker, it would indicate that the voice coil was open or loose from the cone. If point A produces a sound in the speaker, then move the generator probe to point B. If no sound is heard in the speaker, one

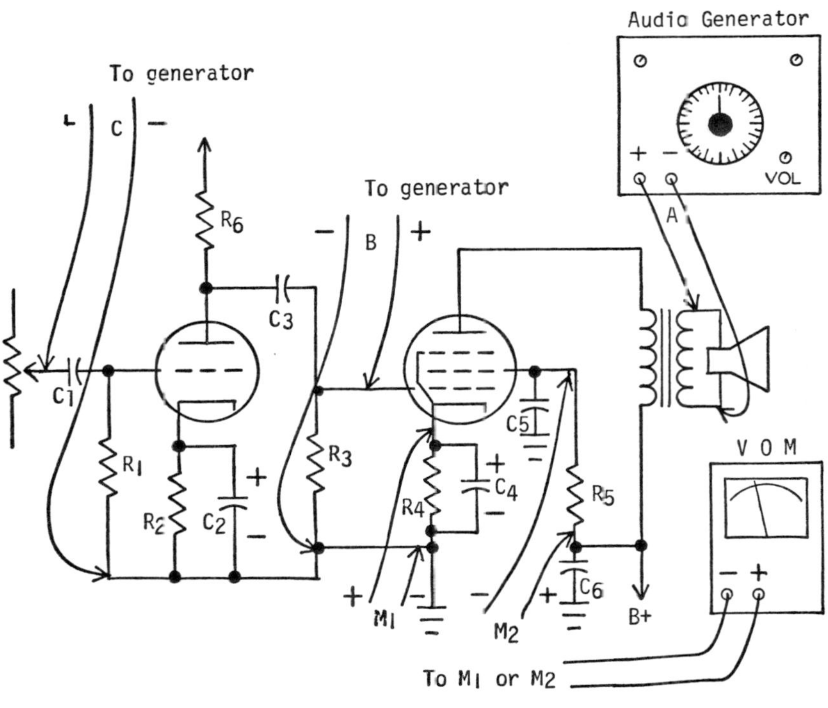

FIGURE 7-8: Audio section of a receiver, showing method of connecting audio generator and volt ohm meter, for troubleshooting.

of the possible causes might be that the cathode resistor R_4 may be open.

To check this, connect the meter as shown by the arrows marked M_1. Read the grid voltage at this point. If the voltage is correct, move the meter leads to the point where the arrows are marked M_2. At this point, you would be reading the drop across the resistor which supplies the screen voltage. Make certain that you observe the polarity of the meter leads as indicated at the lines marked M_2. If capacitor C_5 is open, you will have sound from the speaker, but the volume would be decreased and distorted. If this capacitor were to short to ground, this would place a load on R_5 which would overheat and possibly burn up.

If C_6 were to open up, you would have sound in the speaker, but the audio may break into a high pitch oscillation with high volume. If this capacitor were short to ground, this would put a direct short on the power supply.

If the output stage were found to be operating, then the generator would be placed on point C to inject a signal into the first audio stage. If no sound is heard in the speaker, a possible cause of trouble might be that the cathode resistor R_2 is open; this would cut off the supply of current through the tube. Place the meter leads across this resistor the same way you did across R_4 in the output stage. You will now be measuring the grid bias of this tube. If there is no bias voltage, disconnect one end of the resistor and measure the resistor with an ohmmeter. If the resistor is open, replace it; but if the resistor measures correctly, replace it in the circuit.

Disconnect capacitor C_2 at one end and switch your ohmmeter to its highest resistance scale, placing the leads across the capacitor. Since this is a polarized capacitor, the positive lead of the meter should be connected to the positive end and the negative lead to the negative end of the capacitor. The needle of the meter should go to the very low resistance at first. If the capacitor is shorted, the needle will stay at a very low resistance. If the needle starts returning to the high end of the scale and continues to go toward the high end and goes above 150k ohms, it should be satisfactory.

Replace the capacitor, and with your meter switched to read voltage, place it across R_6, the plate load resistor. Connect the meter leads with the same polarity as you did across R_5. You are now reading the voltage drop across this resistor, which is the supply voltage for the plate of the tube. If the voltage reads correctly, this resistor is all right.

Another possible trouble might be capacitor C_3, which is the coupling capacitor between stages. If this capacitor were open, the signal would not pass through. If it were shorted, this would place a high positive voltage on the grid of the next stage, and you would have seen this while checking the output stage. Check this capacitor to see if a signal will pass, by connecting the generator at the point where the resistor R_6 and C_3 are connected to the plate. If a signal is heard in the speaker, this capacitor is satisfactory. Since there is not a high potential at either end of capacitor C_1, it is not a likely suspect for trouble. The same is true with R_1 and R_3; they are not likely suspects since there is not a high potential at either end of these resistors.

From the second detector stage back to the antenna is the RF section of the receiver. Use the RF generator to inject a signal into these stages. Figure 7-9 shows how to connect the RF generator to the IF and mixer-converter stages to inject a signal. The generator leads are marked with the letter G to indicate the generator, and the test points are marked G_1 and G_2. The ground lead is left connected to the same point for all connections. The first test point is G_1 with the generator leads connected to this point; the generator output should be set for 456 KHz. This is the IF amplifier stage; its frequency is 456 KHz. The generator should be set for modulated signal output. If a signal is heard in the speaker, this stage is satisfactory. The generator leads should be moved to point marked G_2, and since this is the input of the mixer-converter stage, the input signal should be an RF signal corresponding to the dial setting. This should be at some frequency between 550 KHz and 1600 KHz. If a signal is heard in the speaker, the RF section of this stage is satisfactory. Now check the oscillator section of this tube and use the oscilloscope to check the oscillator voltage. The oscilloscope is shown connected between the oscillator grid and ground. The wave form on the oscilloscope should show an RF voltage which is the IF frequency. With the oscilloscope connected as shown, vary the tuning capacitor C_2; if the voltage wave form varies while turning this capacitor, the oscillator section is satisfactory.

Let us go back now and suppose that while checking the IF stage at the input or grid of this stage, you did not hear a signal from the speaker. Now apply the voltage check method in troubleshooting. In the diagram 7-9, a voltmeter is shown; its negative lead is shown connected to ground which can be connected during all of the voltage measurements. The positive lead of the meter is marked with the letter

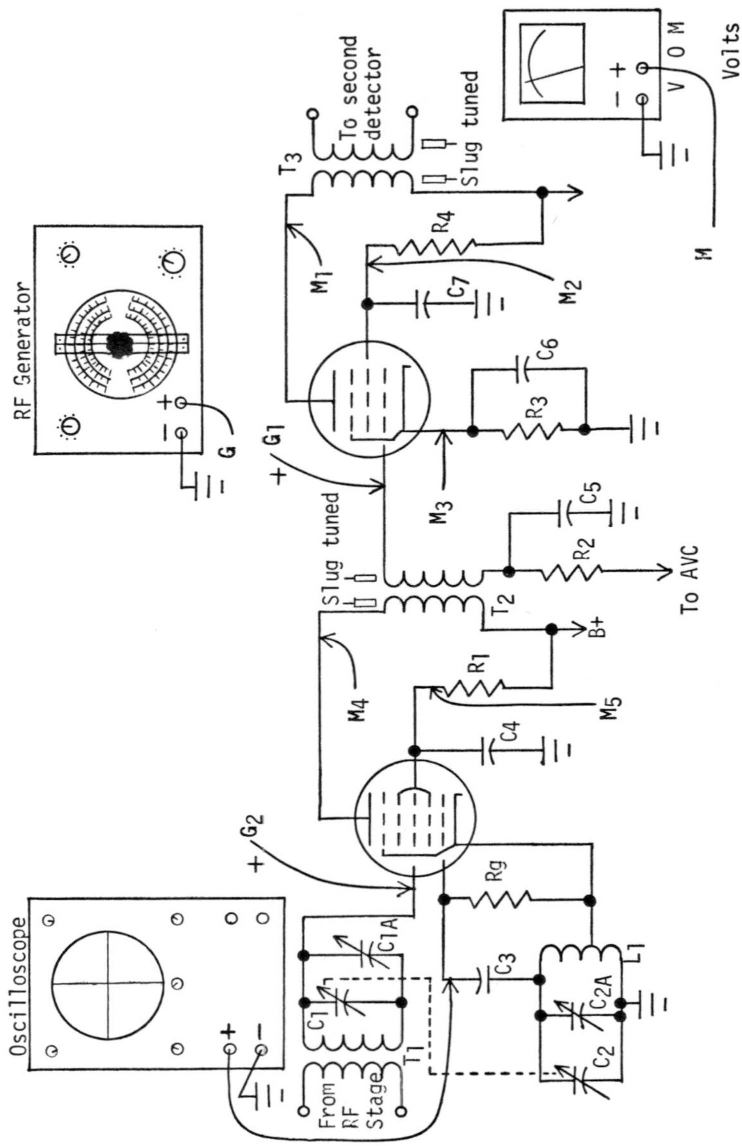

FIGURE 7-9: IF and mixer-converter stages with test equipment connections for troubleshooting.

M, and the test points are marked with the letter M and a number. This is to indicate the sequence at which the various tests should be made.

First connect the meter to M_1; you are measuring the plate voltage on the tube. If the primary of the transformer is open, you will not have voltage on the plate. If this voltage measures satisfactorily, move the meter lead to point M_2; you are measuring the screen voltage. If R_4, the screen resistor, is open you will not have voltage on the screen. If C_7, the screen decoupling capacitor, is open, the gain in the tube will be very low, and the reception will only be a very strong local station. If this capacitor is shorted to ground, it will put a load on the screen resistor, causing it to draw an excessive amount of current that would make it burn up. If the voltage is satisfactory on the screen, move the meter lead to point M_3. You are measuring the grid bias or the voltage drop across the cathode resistor. If this resistor opens up, there will be no plate current and the tube will not operate. If capacitor C_6 shorts to ground, there will be no grid voltage, and the plate of the tube will become overheated since it is drawing an excessive amount of current. If this capacitor is open, there is a loss in gain, and the stage may break into oscillation during strong signals.

Capacitor C_5 and resistor R_2 are parts of the AVC circuit, and since there is no high potential voltage at this point, it is not likely that either of these two would be defective. Now that the amplifier stage checks satisfactorily, change the meter leads to point M_4. You are reading the plate voltage of the mixer-converter stage. If the primary winding of the IF transformer opens up, there will not be plate voltage on the tube. If the voltage is satisfactory here, move your meter leads to point M_5. You are reading the screen voltage of the tube. If R_1 opens up, there will not be screen voltage on the tube. There may have been an excess load on this resistor at some time, causing it to change value. When a carbon resistor changes value due to heating, it always changes to a higher value. With the meter connected to point M_5, if the meter is reading a low voltage, this may be the reason. The resistance has changed to a higher value, and the voltage drop across it will be more. The screen voltage is the difference between the drop across this resistor and the supply voltage. If the capacitor C_4 shorts to ground, this puts an excess load on the resistor R_1 and causes it to burn up. If this same capacitor opens, the gain of the stage decreases due to regeneration between the cathode and screen circuit. If the voltage at point M_5 is satisfactory, this would complete the voltage checks.

The trouble may be the oscillator stage. There are two ways of checking the oscillator stage to observe its operation. One is by using the oscilloscope, and the other is by using the VTVM and checking the grid voltage of the tube. Since you have used the oscilloscope earlier in checking, use the VTVM in this test.

Connect the VTVM negative lead to the point where the positive lead of the oscilloscope is shown. Connect the positive lead to the cathode. You are now reading grid voltage. Rotate the tuning dial; if the reading varies as you change frequency, the oscillator is working.

This completes the troubleshooting procedure through the receiver. You could have reversed this procedure, starting at the atennna and going back through to the speaker. Either way would have been correct. Where you choose to start will depend upon the trouble. By choosing the proper method, you will save time and effort in your troubleshooting. In our troubleshooting method, we did not include the RF stage because, as mentioned earlier, not all receivers have an RF stage. In case the receiver you are working with has an RF stage, the method of checking it will be the same as the method used in the IF stage.

The main thing to remember when troubleshooting any type of electronic equipment is to isolate the trouble to one stage and then work on that stage. Since localizing the trouble is the most time consuming part of the job, use the correct procedure to save time. Installing the new part may take only a few minutes.

ALIGNMENT OF THE RF AND IF STAGES

The alignment of the receiver is not one of the troubleshooting methods, but there are times when it may be necessary to align both the RF and the IF sections of the receiver. In some cases, where the mixer-converter tube is replaced, due to the difference in the interelectrode capacitance of the tube the oscillator or RF section may need alignment. This may be true in the IF section also; therefore, I shall discuss the correct method of aligning both stages.

Figure 7-10 shows the speaker and output transformer with the voice coil lead disconnected and the VTVM connected across the voice coil leads of the output transformer. The meter pointer can be adjusted to the center of the scale; it is acting only as an output indicator. You are not concerned with the value it is reading; only as the

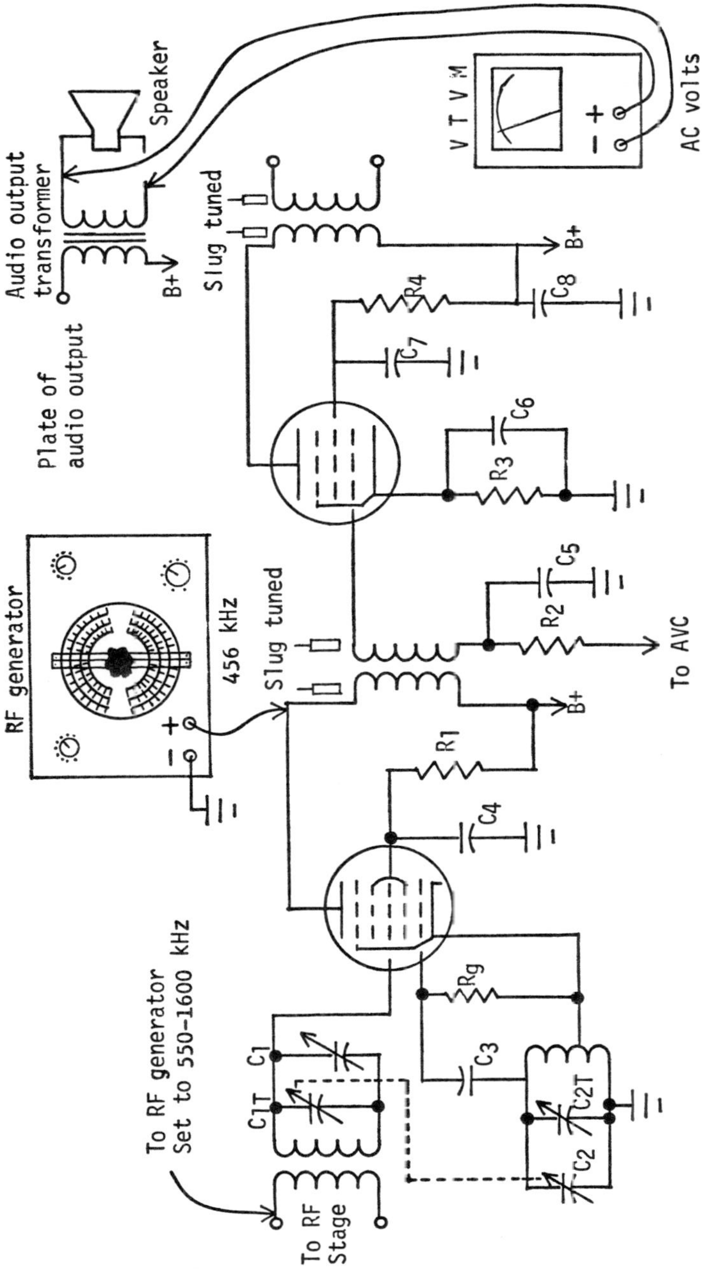

FIGURE 7-10: IF and mixer-converter stages with instruments for alignment.

adjustments are made, does the meter pointer move in an upward movement.

The RF generator should be set at 456 KHz which is the frequency of the IF. This signal can be unmodulated and the RF gain control set to give a deflection on the meter. This control should be kept turned down during the alignment just to the point where you can see the meter deflect as you adjust the tuning. Set the receiver dial to 550 KHz or the variable capacitor fully meshed; attach the generator leads to the point shown in the diagram. Turn the generator on and allow it to warm up for approximately five to ten minutes; then turn the receiver on.

Begin with the secondary of the second IF transformer and adjust the slug in the secondary of the second IF transformer for maximum deflection on the meter. Next, adjust the slug in the primary of the second IF transformer for maximum deflection on the meter. Now, adjust the slug in the secondary of the first IF transformer for maximum deflection on the meter. Adjust the primary of the first IF transformer for maximum deflection on the meter. Return to the secondary and primary of the second IF to see if they may need a little more adjustment, again for maximum deflection on the output meter. Again, throughout the adjustment, keep the RF level of the input signal as low as possible to keep the IF stage from overloading.

In some cases in the alignment it may be necessary to disable the AVC circuit to avoid AVC action that would reduce the gain in the IF stage due to the strong signal being applied to that stage from the generator. This could be done by disconnecting C_1 in Figure 7-6 or by disconnecting R_4 in the same diagram.

Attach the generator leads to the RF section as shown in Figure 7-10. Adjust the generator dial to a setting of 550 KHz. Leave the receiver dial set where you have it, at the low end of the dial which is 550 KHz. Adjust the RF level of the generator until the meter is moving upward from center where you set it at first. Use an insulated adjustment tool for this adjustment. You are going to check the oscillator alignment for proper dial calibration. Notice the meter reading so that you may return to this point if necessary. Locate the trimmer screw on the oscillator variable capacitor; this is the smaller of the two variable capacitors. This screw is marked $C_{2}T$ in Figure 7-10. Place the tool in the screw slot and move it slowly and watch the meter. If the meter starts to fall back toward zero, move the screw in the opposite direction, again watching the meter. If the meter returns to the original point and then starts to fall back toward zero, move it to its original point. Move the generator dial to 1600 KHz, and move the dial of the receiver to 1600 KHz also. Rock the receiver dial back and forth. If the maximum signal falls on 1600 KHz, then the receiver dial is as close to calibration as you will be able to obtain. Move the generator dial to 1000 KHz and the receiver dial to 1000 KHz; you are now in approximately the center of the broadcast band. Locate the trimmer screw on the larger of the two variable capacitors; this is the RF or mixer-converter variable capacitor. This trimmer is marked C_1T in Figure 7-10 and should be adjusted for maximum deflection on the meter.

Disconnect the generator and the meter; solder the wire back to the speaker and the wires or components in the AVC circuit. Turn the receiver on and check the stations on both ends of the dial and the center of the dial. This completes the alignment.

8

Communication—Sending: Frequency Modulation Transmitters

FREQUENCY MODULATION DEFINED

Frequency modulation is defined as follows: modulation in which the instantaneous frequency of the modulated wave differs from the carrier frequency by an amount proportional to the instantaneous value of the modulating wave. The amplitude of the modulated wave is constant. The frequency modulation broadcasting system is practically immune to atmospheric and man-made interference.

The frequency modulation band extends from 88 to 108 MHz, and this band is divided into channels which are 200 KHz wide and are designated by their center frequency. Assigned center frequencies begin at 88.1 MHz and continue in successive steps of 200 KHz up to 107.9 MHz.

We have already learned in our study of AM that the carrier is varied in amplitude by the frequency of the audio, and that this variation in amplitude of the carrier and audio formed the modulation envelope. In the study of FM, just the opposite is true. Instead of the amplitude of the carrier being varied by the audio, the frequency of the carrier is varied by the amplitude of the audio.

Figure 8-1 illustrates the difference between an AM modulating envelope and an FM modulated envelope. Notice in (a) that the amplitude of the carrier is varying, while the frequency remains the same. In (b), the amplitude remains the same but the frequency is varied. Also shown in (b) is the assigned band width of each channel. The dotted line, shown in the center, is the assigned frequency of the station; the station is allowed to deviate this carrier 75 KHz above this frequency

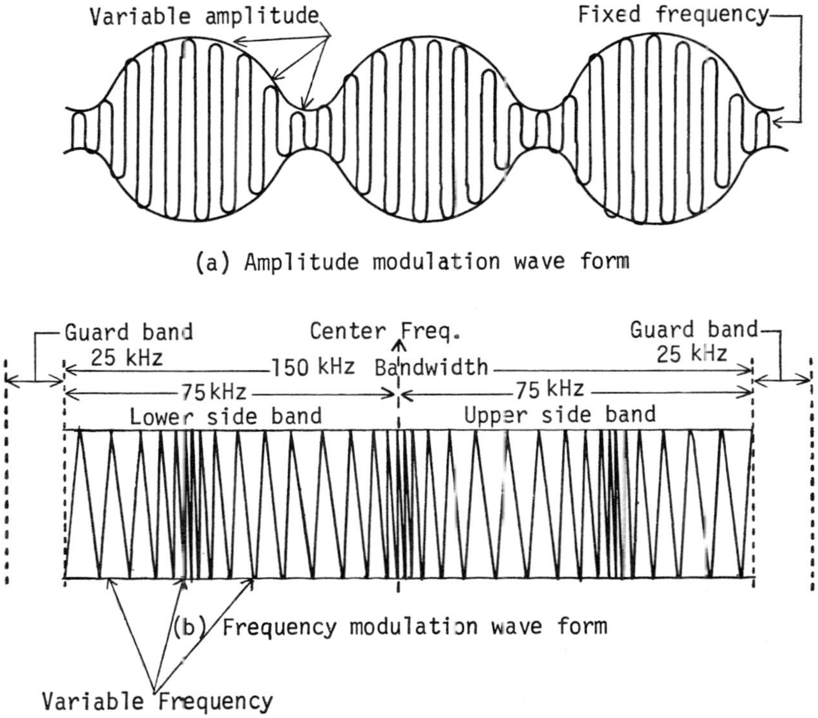

FIGURE 8-1: Comparison between the AM and the FM wave forms. Also shown is the assigned band width of each channel.

and 75 KHz below this frequency. The complete deviation of the carrier allowed for the station is a band width of 150 KHz. This is considered 100% modulation. The FCC (Federal Communication Commision) has allocated 25 KHz above the upper side-band and 25 KHz below the lower side-band as guard bands. These guard bands on either side of each channel protect the adjacent channels from being interfered with, in case some station should overmodulate. To avoid this condition, no two stations are assigned adjacent channels in the same area.

Unlike amplitude modulation, FM has many frequencies appearing in each side-band, the number depending upon the amplitude and frequency of the modulation which deviates the carrier. The highest AF frequency required by an FM station is 15 KHz, and the widest carrier deviation allowed is 75 KHz, so that the ratio of the greatest carrier deviation to the highest modulating frequency would be 75:15. This is called the deviation ratio, and the ratio of maximum carrier deviation to any particular modulating frequency is called the modulation index. We can determine the modulation index by the following formula:

$$\text{Modulation index} = \frac{\text{carrier deviation}}{\text{modulating frequency}}$$

By substituting the highest allowable carrier deviation (75 KHz) and the highest modulating frequency used by FM stations (15 KHz), the modulation index = $\frac{75}{15}$ = 5, which is the maximum for the highest modulating frequency. As the modulating frequency goes lower, the number of modulating index increases. For instance, at a modulating frequency of 100 cycles, the modulation index would be 750:

$$\text{Modulation index} = \frac{75{,}000}{100} = 750$$

The variation of an index of from 5 to 750 appears in both the upper and lower side-bands which make up the assigned 150 KHz band width. Each side-band is deviated by many frequencies between 100 Hz and 15 KHz, which is the audio range normally used in FM broadcasting.

THE FREQUENCY MODULATED TRANSMITTER STAGES

The oscillator stage of a FM transmitter is usually operated in the lower frequencies for greater frequency stability. The oscillator stage is followed by a buffer doubler stage, and each succeeding stage is a doubler or tripler stage until the assigned frequency is obtained in the final output stage. Figure 8-2 illustrates a block diagram of a basic FM transmitter; notice that the modulation is applied to the low level stages of the oscillator. In the AM transmitter, the modulation was applied to the high level stages or the final stages. Since the carrier deviation is a result of the amplitude and frequency of the modulation, this modulation is applied to the stage where the carrier is generated—the oscillator stage.

In the diagram the wave forms are shown as they appear in the various stages of the transmitter. Notice in the audio stages the variable amplitude of the wave form as the different frequencies are developed.

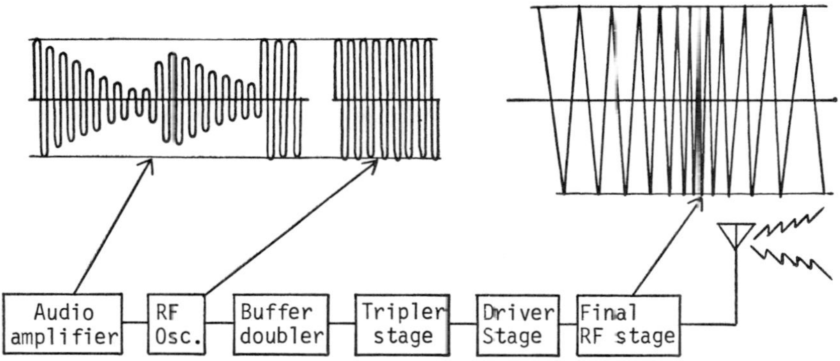

FIGURE 8-2: Block diagram of a basic FM transmitter with wave forms for audio and RF.

It is this variable amplitude caused by the varying frequencies in the audio that deviates the carrier. The wave form from the oscillator circuit is uniform in amplitude as well as in the spacing between each hertz.

Figure 8-3 illustrates just how the carrier is deviated by the amplitude of the modulating audio. In (a) the top wave form shows the

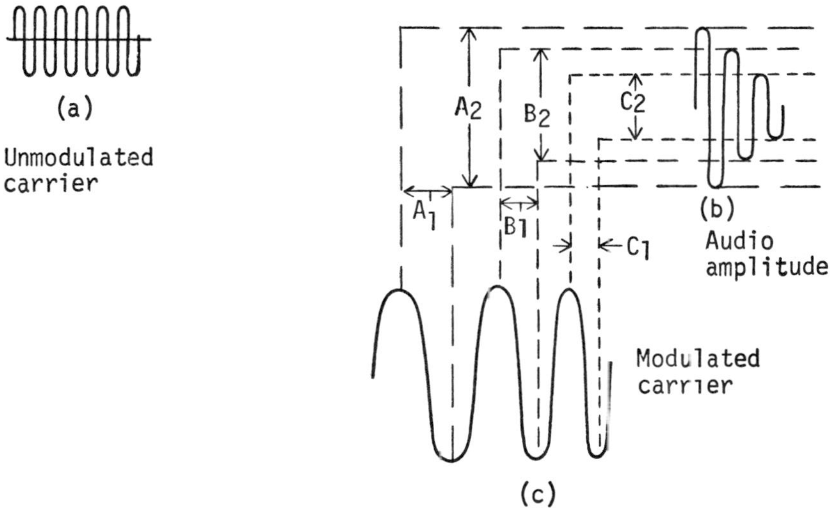

FIGURE 8-3: Illustration of how the strong, medium and weak amplitude audio modulates the FM carrier.

unmodulated carrier, while the wave form at (b) shows the audio wave form. The wave form in (c) is the modulated carrier wave form. Notice in (c) that the width of the wave form A_1 was caused by the amplitude of the wave form A_2 of the audio in (b). B_1 in (c) was caused by the audio amplitude of B_2 in (b). The width of the modulated carrier C_1 in (c) was caused by the amplitude of the audio C_2 in (b). Notice how wide the carrier is expanded with the strong signal at A_1, as compared to the width of the carrier C_1, caused by the weak signal C_2 in the audio wave form.

In Figure 8-3 we have broken the audio wave form and the modulated carrier wave form down into segments, to show how the carrier is shifted in frequency during modulation. This frequency shift was caused by the amplitude of the strong, medium, and weak audio applied to the oscillator stage of the FM transmitter.

FREQUENCY MODULATION—STEREO MULTIPLEX

Most FM broadcasting stations have been using the stereo system of broadcasting, and it is only recently that the FCC authorized a system of stereo called stereo multiplex. This system can be received on the regular FM stereo receiver, without any changes, and many of the FM broadcast stations are using this system today.

First let us define these two words, *stereo* and *multiplex*. Stereo may be defined as "pertaining to three-dimensional pickup or reproduction of sound, as achieved by using two or more separate audio channels." Multiplex may be defined as "the simultaneous transmission of two or more programs or signals over a single RF channel, such as by time division, frequency division, or phase division."

In listening to stereo, someone once commented, "It is like hearing from both ears at the same time." In referring to stereo, we do use the terms left channel and right channel, and they do give the listener a two-dimensional sound effect, making him feel that he is actually there where the sound is originating. What he hears on his left from the speaker, he would actually be hearing, if he were there, on his left side. What he hears on the right side speaker he would hear, if he were there, on his right side. In listening to a choir from the studio, the effect is accomplished by placing one microphone on the left side of the choir and one on the right side. These are fed into two separate channels and transmitted separately. The receiver detects them separately, and from here they are fed into two separate sound channels. The sound is heard by the listener as two separate signals, just as they were produced in the studio.

In the definition of multiplex, we mentioned that the system was accomplished by time division, frequency division, or by phase division. FM stereo multiplex transmission uses the phase division and frequency division. The left and right channels are fed into a mixer. The output covers the complete audio range from 50 Hz to 15 KHz. This signal is fed into a reactance modulator stage. At the same time the right channel is fed into a phase inverter stage which shifts the signal 180°. This signal, along with the left channel, is fed into another mixer stage. The output of this stage is fed into a balance modulator. A 38 KHz carrier (which is the second harmonic of a 19 KHz pilot sub-carrier oscillator) is fed into a balance modulator. Here the combined audio from the mixer, which is 50 Hz to 15 KHz, amplitude modulates the 38 KHz carrier in the balance modulator. The result here is that the 38 KHz carrier is balanced out, leaving a double side-band with a frequency range from 23 to 53 KHz. This signal now is fed to the reactance tube modulator along with the original 50 Hz to 15 KHz signal. These combined signals now frequency modulate the carrier of the FM transmitter.

9

Communications—Receiving: Frequency Modulation Receivers

SIMILARITY AND DIFFERENCE BETWEEN AM AND FM RECEIVERS

Receivers used for the reception of FM are very similar, stage for stage, to those of the AM receiver, from the RF stage back to the second detector stage.

Like the AM receiver, the FM receiver may or may not use an RF stage. You have learned by theory that the use of an RF stage in any type of receiving equipment will improve the sensitivity and selectivity; thus in the better FM receivers, an RF stage is used.

Following the RF stage is the converter or mixer stage where the principle of heterodyning is used. In the AM receiver, a special tube was designed (pentagrid converter) in which the detector and oscillator stages were combined. In the FM receiver, the frequency at which this stage must operate is much higher, so these two stages are separated. The reason here is the higher in frequency an oscillator circuit must operate, the more stable the circuit must be. This is especially true in the transistor type oscillator circuit; at higher frequencies, more heat is generated. Since the transistor is very critical in operation during temperature variations, as a separate oscillator circuit, greater stability can be maintained.

Following the mixer stage is the IF (intermediate frequency) stage. Some FM receivers use two IF stages; this is found in the more expensive receivers. Following the last IF stage, the receiver differs from the AM receiver. Instead of the second detector stage, as in the AM receiver, you will find a limiter stage and a discriminator stage. In

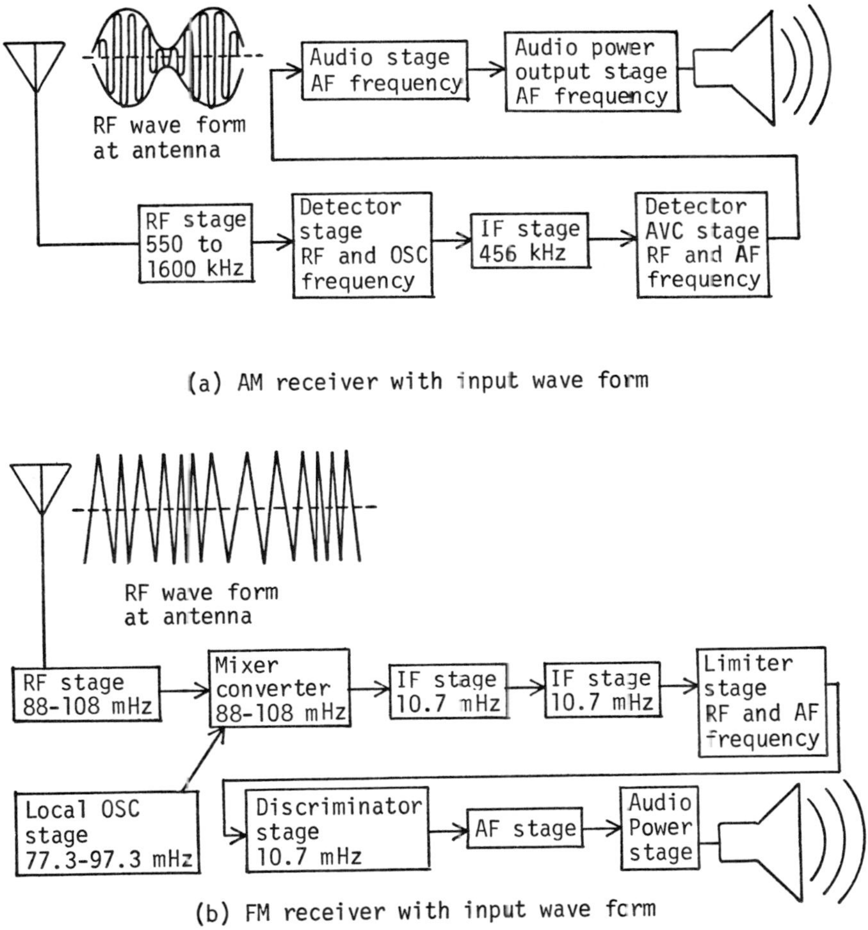

(a) AM receiver with input wave form

(b) FM receiver with input wave form

FIGURE 9-1: Comparison of AM and FM receivers.

this stage the RF carrier and intelligence are separated, and from this stage the audio is applied to the first audio stage and then to the last audio stage.

Figure 9-1 illustrates, by block diagram, the comparison of the AM receiver and the FM receiver. In (a) is shown the AM receiver with the input wave form at the antenna of the receiver. In (b) is shown the FM receiver and input wave form at the antenna and frequency of each stage.

The principal differences between the AM and FM receivers are the frequency which each must be capable of tuning and the type of modulation or method of modulating the carrier (amplitude or frequency).

RECEIVER STAGES—RF AMPLIFIER

The FM band covers a 20 MHz width in the radio spectrum. The low end of the band is 88 MHz, and the high end of the band is 108 MHz. Each channel is 150 KHz wide for 100% modulation, which is the allowable band width for each station. At the lower and high ends of this channel, the FCC (Federal Communications Commission) has set aside 25 KHz as a guard band. This makes each channel 200 KHz wide, as compared to the 10 KHz width of each channel in the AM band.

If you were to inspect visually the top and underside of the chassis of the FM receiver, some of the things you would find different from the AM receiver are as follows:

1. In the RF stage, you might notice that all leads are kept as short as possible, especially those carrying the RF signal as well as the ground leads. At this frequency, the least amount of extra wire left on the leads of some components might introduce stray capacitance or inductance. The use of any excess amount of conductor wire between components might have the same effect.

2. You would notice that the inductors or coils are much smaller and have only a few turns which are wound with much heavier wire. At this frequency less inductance is needed; hence coils can be made smaller. Also, you would notice that the variable capacitor is smaller in physical size and has fewer plates. At this frequency less capacitance is needed to resonate the lesser inductance, and smaller size capacitors are used for tuning.

3. In some of the better receivers you may notice small RF choke coils in the B+ leads, with bypass capacitors on the power side of the choke. This keeps the AC signal from going back into the power supply and returning to the stage input as an out-of-phase voltage that would cancel out part of the signal.

Refer to Figure 9-2, which illustrates an RF amplifier that might be used in the first stage of an FM receiver. Since the modern receivers are all solid state, we are showing the transistor amplifier and the complete solid state FM receiver.

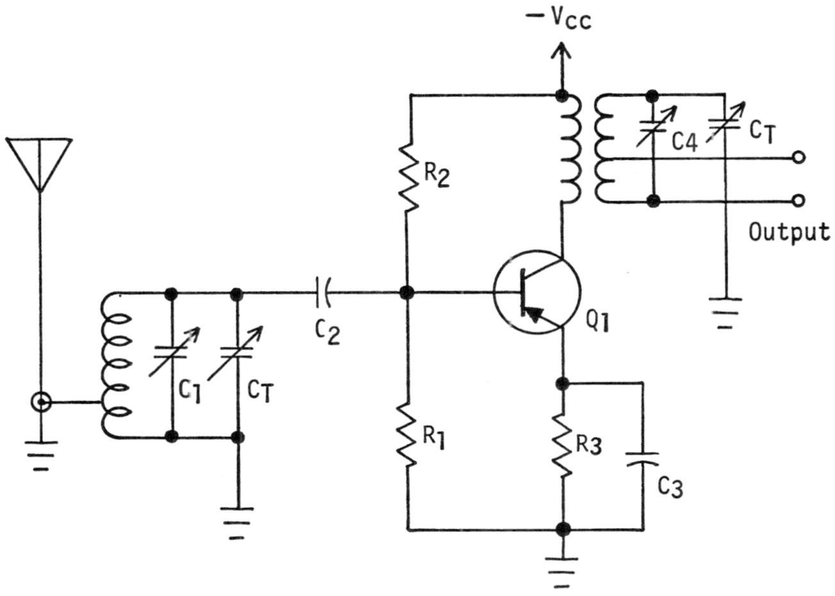

FIGURE 9-2: One type of RF amplifier that might be used in a solid state FM receiver.

The antenna is connected to an antenna coil which is tuned by C_1, a variable capacitor, and C_t is the trimmer capacitor which is mounted on the side of the variable C_1. Capacitor C_2 couples the RF signal to the base of the transistor. Resistor R_2 is the base bias resistor, and base current flows down through this resistor, causing a voltage drop. The difference between the supply voltage and this drop is the base bias voltage or the forward bias. The current passes through the emitter resistor R_3 and back to ground. In this circuit, the positive side of the supply is the common tie point. Resistor R_1 is the stabilizing resistor. When collector current increases during a rise in temperature in the collector region, a larger voltage is developed across this resistor, which reverse biases the base emitter junction and cuts the transistor off. C_3 is the emitter bypass capacitor which bypasses the AC signal around the emitter voltage supply (R_3). The transformer in the collector stage is the RF transformer which couples the signal to the mixer stage.

Figure 9-3 shows the mixer oscillator stages. Q_2 with its associated components is the mixer stage, while Q_3 with its associated components is the oscillator stage. The RF signal from the RF amp-

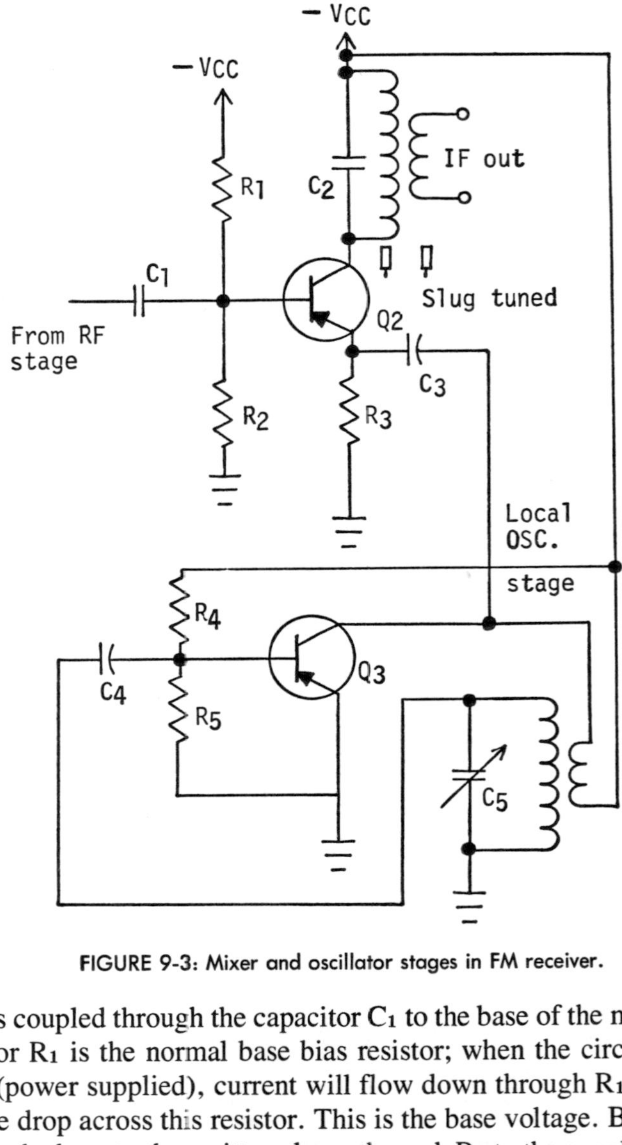

FIGURE 9-3: Mixer and oscillator stages in FM receiver.

lifier is coupled through the capacitor C_1 to the base of the mixer stage. Resistor R_1 is the normal base bias resistor; when the circuit is energized (power supplied), current will flow down through R_1, causing a voltage drop across this resistor. This is the base voltage. Base current is from the base to the emitter, down through R_3 to the supply. Resistor R_2 is a bias stabilizing resistor. When current increases in the collector region due to an increase in temperature in the collector, a reverse current flows from the collector through the base and down through this resistor. This increases the base voltage which reverse biases the base and cuts the transistor off. The transformer in the collector stage is the IF transformer, which couples the signal to the IF stage. The lower circuit Q_3 is the oscillator circuit connected as a common emitter amplifier. The voltage divider network of R_4 and R_5 is the bias and bias stabilizing resistors. C_5 is a variable capacitor across the inductor, which is a resonent circuit, tuning from 77.3 MHz to 97.3 MHz, or the oscillator frequency. The feedback voltage is coupled to the base by capacitor C_4, and the collector is connected to the secondary of the inductor. The oscillator voltage from the collector circuit is coupled to the mixer emitter through capacitor C_3.

The mixer stage is a heterodyning action in the same manner as the AM receiver, only at a different frequency. The FM band extends from 88 MHz to 108 MHz, and one of these frequencies is produced at the base of the mixer stage. The oscillator of the receiver will tune from 77.3 MHz to 97.3 MHz, and one of these frequencies is produced at the emitter of the mixer. The difference between the RF and the oscillator voltages in the mixer stage then produces the IF frequency at the collector of the mixer stage, which is 10.7 MHz. The collector of the mixer stage is coupled to the primary of the IF transformer, and this voltage is coupled to the IF amplifier by the secondary of the transformer.

To better understand this heterodyning action, assume you have tuned the FM dial to the lower end, or 88 MHz. This is the RF signal. At the same time the dial is tuned to this frequency, the oscillator is tuned to 77.3 MHz. An RF signal will appear at the base of 88 MHz and an oscillator signal at the emitter of 77.3 MHz. When they are combined in the mixer stage, their difference will be 10.7 MHz, which is the signal appearing at the collector of the IF transformer. Now, suppose you tuned the dial to the high end, which is 108 MHz. You have tuned the oscillator to 97.3 MHz. These two signals are appearing in the mixer stage, and their difference is 10.7 MHz. This is the IF frequency which is appearing in the collector and IF transformer.

With any tuning of the dial in the FM band, the oscillator will always be 10.7 MHz below that frequency. Assume you have tuned to the center of the FM band, which is 93 MHz. To what frequency is the oscillator tuned? Since the oscillator is always 10.7 MHz below the RF carrier, it is tuned to 93 minus 10.7 or 82.3 MHz.

Since FM receivers are using transistors, greater care must be used in designing the oscillator circuit than in the AM receiver. Transistors are normally critical under varying temperatures; as a result special precautions must be taken to prevent frequency drift. The oscillator circuit is placed in a position where it will not be affected by components that are dissipating any amount of heat. Capacitors used in the circuit are usually made of material with low temperature coefficient. Negative temperature coefficient capacitors are sometimes used in the oscillator circuit. An AFC (automatic frequency control) circuit is used to aid in stabilizing oscillator frequency.

The mixer stage is followed by the IF stage; here, the frequency is 10.7 MHz, and the band width is 150 KHz.

Refer to Figure 9-4; this is a typical IF amplifier stage of an FM receiver. T_1 is the first IF transformer; the RF signal from the mixer stage is coupled to the IF by this transformer.

Resistor R_1 couples the AGC (automatic gain control) from the AGC bus line; C_1 is a filter capacitor in the AGC circuit. The signal is coupled to the base from a tap on the secondary of the IF transformer, while C_3 is a neutralization capacitor between the collector and base to prevent oscillation from occurring in the IF stage. Resistor R_2 is the emitter resistor; when the transistor is conducting, current flows down through the emitter and through this resistor, causing a voltage drop. This is the emitter voltage or forward bias for the transistor. Capacitor C_2 is the emitter bypass which bypasses the AC signal around the bias supply, which is the voltage developed across R_2. Resistor R_3 is a low

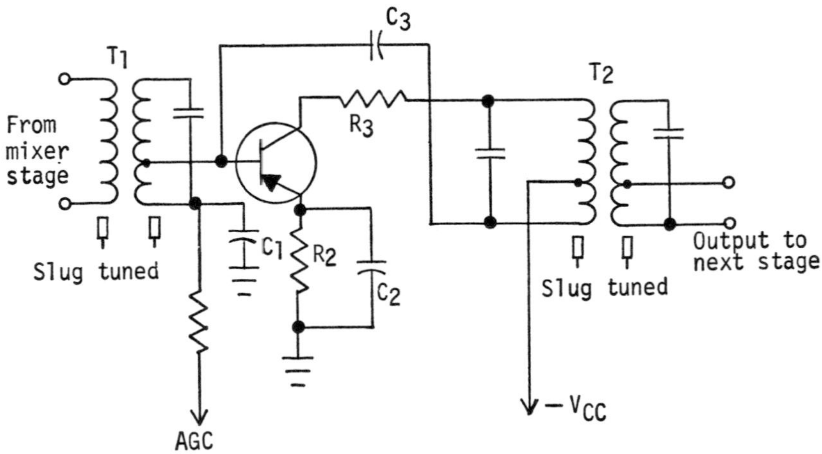

FIGURE 9-4: IF amplifier stage in FM receiver.

value of resistance and is used in some receivers as a feedback resistor. Some transistor amplifier circuits, during very strong signal input, will cause the collector base to become forward biased and cause a feedback voltage to regenerate between the output and input. This causes an oscillation in the stage; so, by installing a small value of resistance in series with the collector, this will limit the signal feedback and prevent regeneration from occurring. Transformer T_2 couples the signal from the collector of the IF amplifier to the next stage.

If the receiver should use two IF stages, the second will be the same as the first. The last IF stage is then coupled to a limiter stage; this stage along with the discriminator stage would be similar to the second detector stage in the AM receiver.

Figure 9-5 shows a limiter stage that might be used in a FM receiver. The transformer in the base circuit is the last IF transformer which couples the signal to a common emitter amplifier. Capacitor C_3 couples the signal to diode D_1, which is a part of a half-wave rectifier circuit. This is the AGC circuit; here, the signal is rectified, filtered, and becomes a DC voltage fed back to some of the preceding stages as

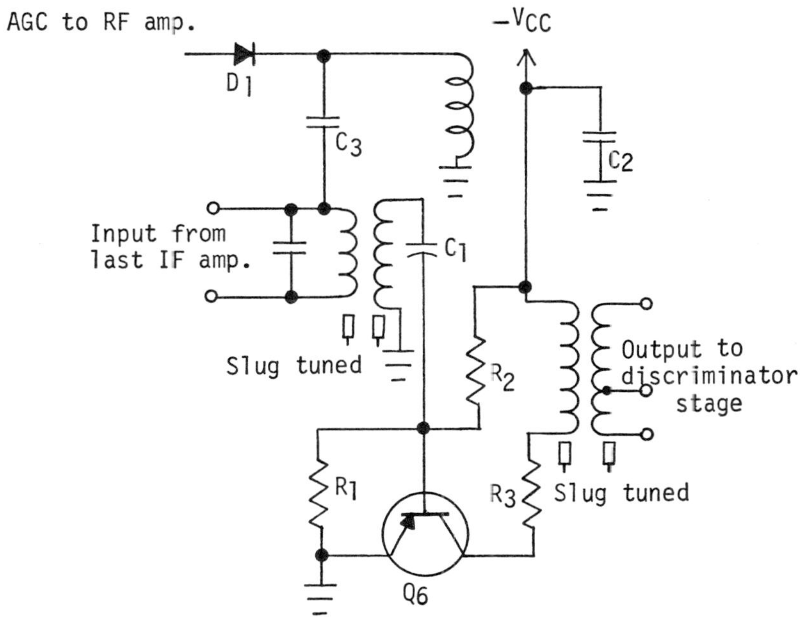

FIGURE 9-5: Limiter stage in FM receiver.

a bias supply. The amplitude of this voltage varies with the strength of the incoming signal, and when it is fed back to some of the preceding stages as bias, it is automatically controlling the gain of these stages. Thus, the over-all gain of the receiver is controlled automatically, as the incoming signal varies in strength.

The secondary winding of the transformer is coupled to the base of Q_6 by C_1; resistor R_1 is a base bias stabilizing resistor. Resistor R_2 is the base bias resistor; the voltage drop across this resistor supplies the base bias. The output of Q_6 is coupled to the primary of the discriminator stage. Capacitor C_2 is a decoupling capacitor which bypasses the AC signal to ground; this is necessary so that the signal will not feed back into the supply.

As I mentioned earlier in this chapter, the main difference between the AM and FM receiver is the manner by which the intelligence is removed from the carrier. In the AM receiver the intelligence was removed from the carrier by detection; in the FM receiver, this is accomplished by the discriminator circuit. The discriminator is a circuit that converts frequency deviations from a carrier frequency into corresponding amplitude variations by removing the amplitude modulation from the frequency modulated carrier.

Figure 9-6 shows a Foster-Seeley discriminator circuit, which is one type of discriminator circuit that might be used in an FM receiver. This circuit is more widely used in modern FM receivers and is sometimes referred to as the phase shift circuit. The circuit resembles a full-wave rectifier circuit. However, the results in the output are a bit different. The transformer couples the signal from the limiter circuit; the primary and secondary are tuned to the center frequency which is 10.7 MHz. The RFC, coupled from the center of the secondary winding of the transformer, supplies a reference voltage for the balanced diodes (CR_1 and CR_2).

A voltage in the primary of the transformer is induced into the secondary winding; when this voltage is equal at both diodes, the rectified voltage appearing across R_1 will equal the rectified voltage appearing across R_2. Since the primary and secondary of the transformer are both tuned to the resonant frequency (10.7 MHz), the voltage appearing in the secondary is 90° out of phase with that in the primary. The secondary is center taped; therefore, the voltage appearing in the upper winding is 90° out of phase with that in the lower winding. The two diodes (CR_1 and CR_2) rectify this voltage, and it appears across the two load resistors (R_1 and R_2) as an equal and

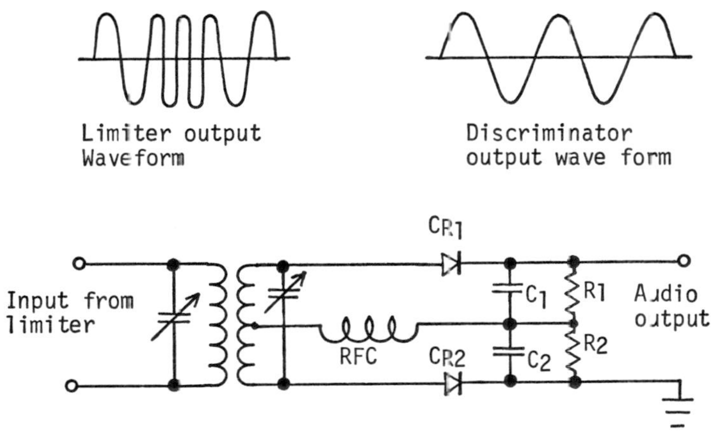

FIGURE 9-6: Foster-Seeley discriminator circuit.

out-of-phase series and opposing voltage. They cancel one another, and the result is a zero voltage at the center of the two balanced resistors. When the incoming signal at the primary of the transformer is above the center frequency, the induced voltage in the secondary is less than 90° out of phase with the primary. During this condition, the secondary is tuned below the signal frequency; the voltage in the upper secondary winding increases because it becomes closer to being in phase. At the same time, the voltage in the lower secondary winding decreases because it is more out of phase with the upper voltage. The voltage at CR_1 is greater than that at CR_2, and a positive output voltage appears across R_1. If the incoming signal in the primary is below the center frequency, the voltage induced into the secondary will be more than 90° out of phase. Under these conditions, the voltage in the lower secondary winding will be greater than that in the upper winding. The rectified voltage across R_2 and the output voltage will be negative. The wave form above the transformer is that of the primary winding voltage; the wave form above the output is that appearing at the output of the discriminator circuit.

One other type of detector circuit quite popular in FM receivers is the ratio detector shown in Figure 9-7. This circuit, which does not require a limiter stage, resembles the discriminator circuit. However, notice that diode CR_1 is reversed to that of CR_2; this makes the secondary winding of the transformer and the two diodes in series. Capacitor

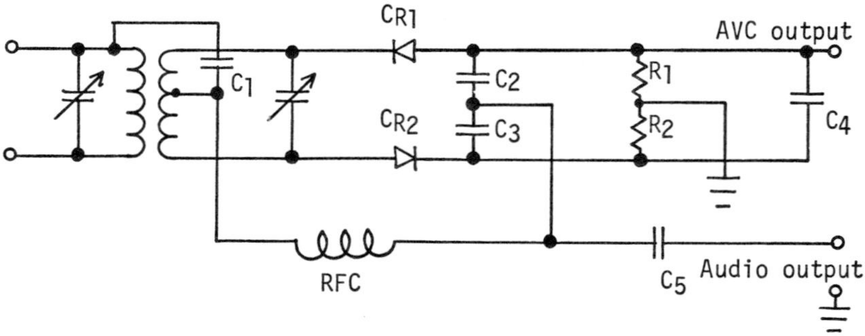

FIGURE 9-7: Ratio detector used in some FM receivers.

C_2 and C_3 are also in series, so that a voltage appearing across these two capacitors will be aiding rather than opposing as in the discriminator circuit. You will also notice that at the point where this series aiding voltage combines (junction of C_2 and C_3), the audio is removed.

The primary and secondary windings of the transformer are tuned to the center frequency (10.7 MHz), and the voltage developed in the secondary will be equal on both sides of the center tap. Since the two diodes are in series, they will conduct at the same time on both alternations of the incoming signal. Thus, the DC voltage developed across the two capacitors (C_2 and C_3) will be equal, and since R_1 and R_2 are shunted across these two capacitors, the voltage developed across them will be equal. Therefore, the voltage at the junction of C_2 and C_3 will be zero as long as the incoming signal remains at the center frequency. Capacitor C_4 is shunted across C_2, C_3, R_1, and R_3. During any amplitude variations in the incoming signal, a voltage is developed across the two resistors and C_4. C_4 is sometimes referred to as a stablizing capacitor because the voltage developed across it is the AVC voltage that is fed back to one of the IF amplifier stages controlling the gain of the receiver.

When the incoming signal is deviated higher in frequency, the DC voltage developed across C_2 will be greater than that across C_3. The audio voltage in the output will have a positive polarity. The incoming voltage in the output will have a negative polarity. Regardless of the deviation of the FM signal above or below the center frequency, the voltage developed across the two capacitors (C_2 and C_3) will always equal the center frequency value. The FM signal deviated at an audio rate will produce a varying DC voltage across these two capacitors.

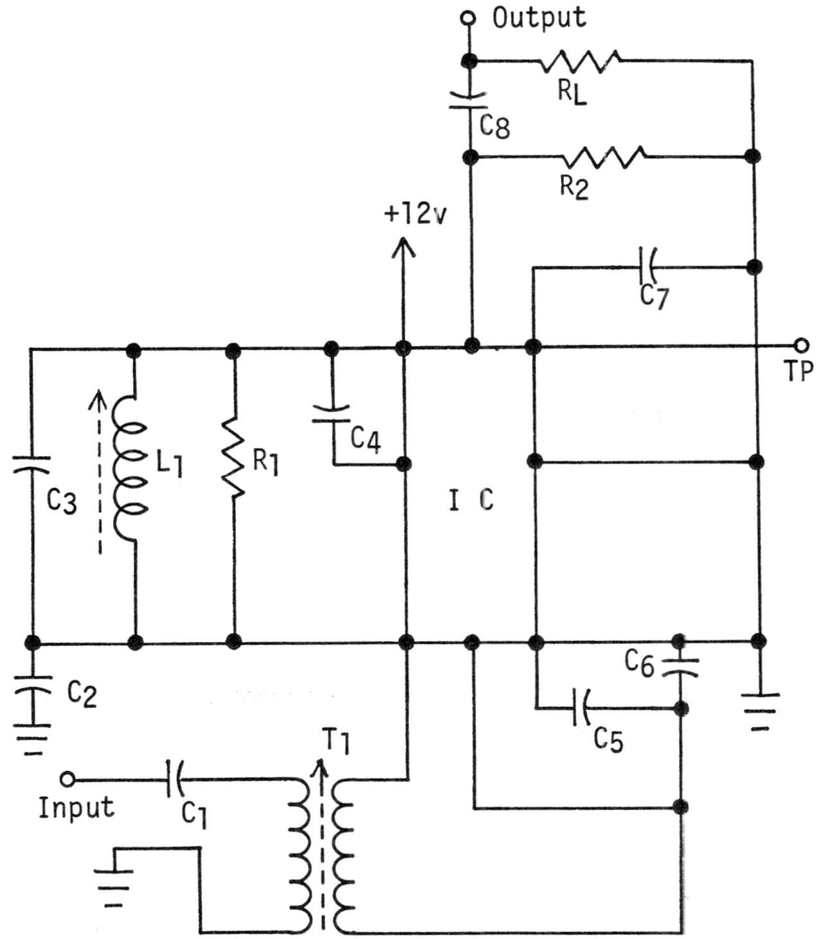

FIGURE 9-8: An IC chip and exterior parts that may be used in place of limiter-discriminator in FM receiver.

One of the main differences between the ratio detector and the discriminator circuit is that in the ratio detector the two diodes are in series and conduct as though only one diode was in the circuit, while in the discriminator the diodes conduct separately, so that the audio output from the ratio detector is half that produced in the output of the discriminator.

Many changes have been made in the AM and FM receivers in the past few years. The PC (printed circuit) board, new transistors improved in both size and efficiency, and recently, the IC (integrated circuit) chips have been introduced. All of these have led to an improvement in performance as well as a reduction in the physical size of the receiver. We have gone from the discrete circuit (made up of distinct parts, such as tubes, transistors, resistors, capacitors, and coils) laid out on the chassis, to one or two IC chips that have all of these parts enclosed in one chip.

Refer to Figure 9-8. This circuit illustrates one IC chip and a few external parts. This replaces the limiter and discriminator circuits in a modern FM receiver. Capacitor C_1 is the regular coupling capacitor, and T_1 is the receiver's last IF transformer. This circuit is much easier to adjust than the two separate circuits. Just apply a 10.7 MHz signal at the point where the transformer connects to the chip and adjust L_1 for maximum audio output at capacitor C_8. There is also a TP (test point) where you may check the output for maximum RF output. This is only one of the many IC circuits used in today's FM receiver.

The audio stages in the FM receiver are no different from those used in the AM receiver. The number of stages and the type of circuit would depend upon the size of the receiver and whether or not it is stereo. Also, in a small personal-size receiver, the audio circuits would not have to produce as much as one of the large cabinet models.

Figure 9-9 illustrates a typical audio section of an FM receiver; this circuit compares very closely to that used in any AM receiver. However, the sound produced in the FM is high fidelity because of the band width used in FM, and the component parts used in the FM sound system of most receivers would have to be of different values and of better quality than those used in the AM receivers.

Refer to Figure 9-9. The first stage is the first audio stage or the driver. C_1 is the coupling capacitor, coupling the audio to the base of the common emitter amplifier. R_1 is the base bias resistor; R_2 is used for bias stabilization; R_3 is the emitter resistor, also for protective bias.

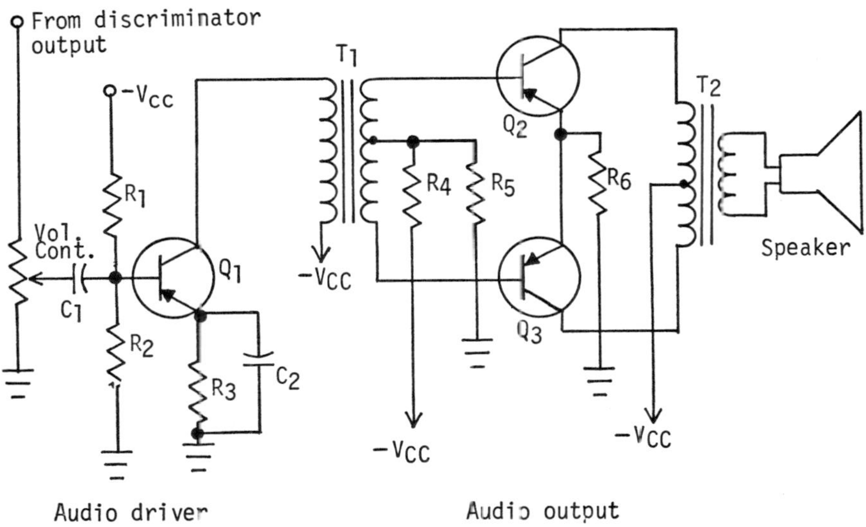

FIGURE 9-9: Audio driver and output that might be used in FM receiver.

C_2 is the emitter bypass capacitor. The signal from the collector is coupled to a driver transformer which has a tapped secondary to couple the signal to a push-pull stage. Q_2 and Q_3 is the push-pull stage; resistor R_4 is the bias resistor, while R_5 is the bias stabilization resistor. Resistor R_6 is the emitter resistor. The collectors of both transistors are coupled to the output transformer and then to the speaker.

ALIGNMENT AND TROUBLESHOOTING YOUR FM RECEIVER

The RF and IF sections of the FM receiver are aligned very similarly to that of the AM receiver. The main thing to remember is that your RF signal generator must tune the FM band 88 to 108 MHz. To align the IF section, the generator must be set at 10.7 MHz.

When we discussed the alignment of the AM receiver, we aligned the IF first and then went to the RF section. In aligning the FM receiver, because of the manner in which the intelligence is detected you will start in the discriminator stage first. Two types of detection have been discussed: (1) the limiter-discriminator; (2) the ratio detector. I shall describe the limiter-discriminator first.

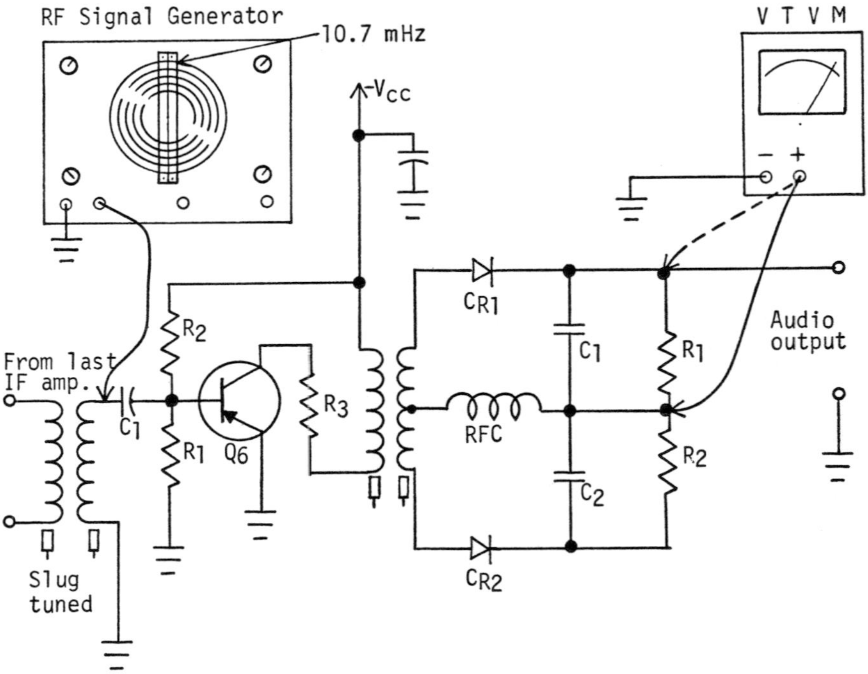

FIGURE 9-10: Aligning discriminator circuit.

Figure 9-10 shows the limiter-discriminator circuit with the RF signal generator tuned to 10.7 MHz, unmodulated and connected to the base of the limiter stage. For the first adjustment, a VTVM is shown connected by the solid line to the junction of the load resistor. Adjust the output of the generator for a deflection on the meter. If you do not obtain a deflection on the meter at first, leave the RF turned up about 1/3 to 1/2, and adjust the secondary of the transformer until you do get a voltage indication on the meter. Tune the primary of the transformer for maximum voltage on the meter. If the meter goes off scale while adjusting this winding, turn the RF gain control down on the generator, and continue to tune this winding until you have maximum voltage indicated on the meter. Leave the generator connected as you have it, and move the meter lead to the point as indicated by the dotted line leading from the meter.

Since one of these resistors is connected to ground, your meter is now across both resistors. Tune the secondary of the transformer until the meter is reading zero. The voltage should be equal across both resistors. To check your alignment, turn the generator 100 KHz above and 100 KHz below the center frequency 10.7 MHz. The voltage drop across each resistor should be equal but opposite in polarity across the two resistors. If this should not occur, go back and check your alignment from the first.

Consider now the alignment of a receiver using a ratio detector. Figure 9-11 shows a ratio detector stage; you will use a RF sweep generator and an oscilloscope to align this stage. You could use the same instruments as before, but this method is probably more accurate and sometimes quicker.

Connect the sweep generator as shown. Apply RF output to the base of the last IF stage. Connect the oscilloscope to the junction of the capacitors C_2 and C_3. Connect the sweep generator output to the horizontal input of the oscilloscope and the sync from the generator to the sync input of the oscilloscope. Set the sweep deviation of the generator to 200 KHz. This will sweep 100 KHz above and 100 KHz below the center frequency 10.7 MHz. You should be viewing an S-shaped curve on the face of the oscilloscope. The vertical portion of this trace shows the varying amplitude of the audio output signal and the horizontal from zero or the center; it also shows the frequency deviation caused by the modulating signal. Adjust the primary of the ratio transformer; this should increase or decrease the vertical, but it should be adjusted for maximum amplitude. Adjust the secondary of the transformer for maximum linearity by watching the horizontal line on the oscilloscope. While going through this alignment, if the S-curve becomes distorted at any time stop and recheck your connections. Make certain the grounding is secure on both the oscilloscope and the generator. This completes the alignment in the detector portion of the FM receiver.

Figure 9-12 shows how you should connect your instruments to align the IF section of the receiver. You will use the same instruments as in the first alignment, the VTVM and the RF signal generator. The diagram shows the last IF stage and the limiter. The VTVM is shown connected to the base of the limiter stage and the negative lead connected to the common tie point which is the positive side of the supply.

First the meter is set on the 15-volt scale. If you do not have a reading after the signal has been applied, move the setting to the next lower range. The signal generator is shown connected to the base of the IF stage, the ground connected to the common tie point. The generator

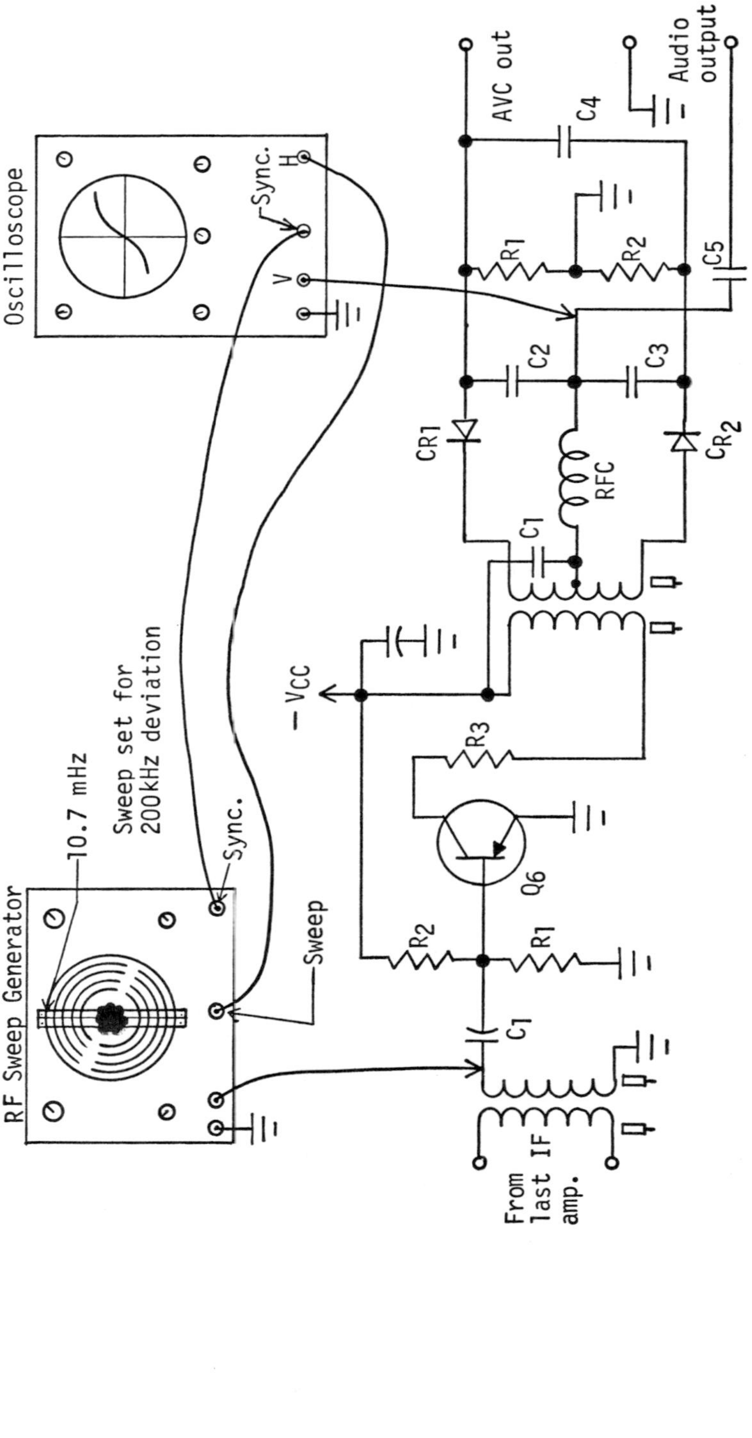

FIGURE 9-11: Aligning ratio detector.

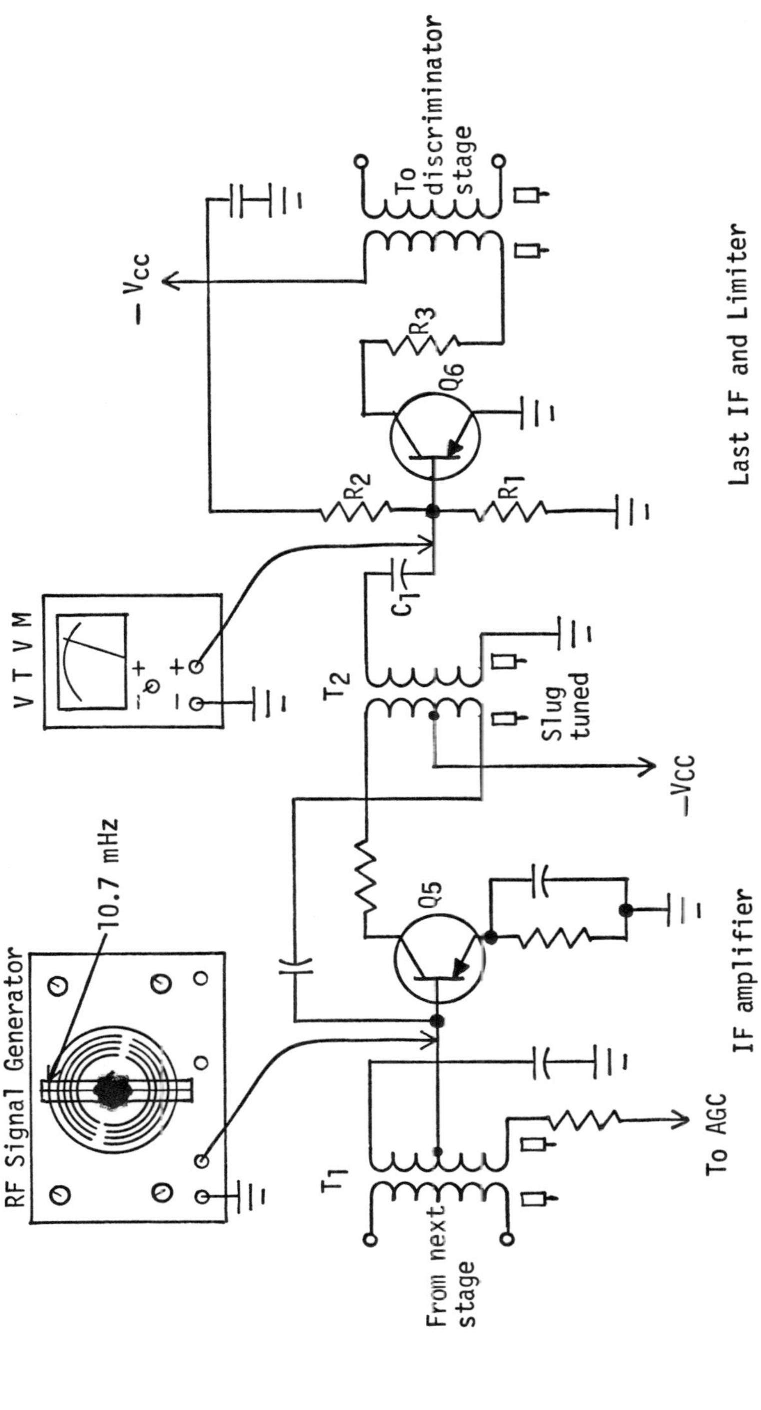

FIGURE 9-12: IF and limiter stages alignment.

is set at 10.7 MHz. The RF gain control is adjusted to give a reading on the meter. Adjust the primary of T_2; this is the side that is connected to the collector of Q_5. Adjust this winding for maximum deflection on the meter. If the meter should go off scale during this adjustment, reduce the gain on the generator and continue to adjust until the meter starts falling down scale. Adjust the secondary of T_2 for maximum deflection on the meter.

If the meter should go off scale, reduce the RF gain control on the generator. Do not reduce the gain until you do not have a signal from the generator, but only enough so that you will not drive the stage with too much signal. With the generator set to high, you are applying more signal than would usually be applied under normal operating conditions. If you have reduced the gain on the generator as much as you can and the meter still reads too high, turn the meter up to the next scale. If the receiver has two IF stages, leave the meter connected at this point and move the generator back to the base of the next stage.

From here back to the antenna, you can leave the meter connected where it is; when you reach the detector stage, you will set the generator to the lower end of the FM band 88 MHz. From here on, the alignment will be the same as it was in the AM receiver. Always keep the gain on the generator down as low as possible and still have a deflection on the meter.

TROUBLESHOOTING

The procedure used in checking a defective FM receiver is very similiar to that used in checking the AM receiver. Due to the frequency and band width used in FM, some of the test equipment will differ. For instance, the RF signal generator you use both for alignment and for signal injection must be capable of tuning the FM band. Most modern FM receivers use either transistor or IC chips. DO NOT ATTEMPT TO USE A VOM AS PART OF YOUR TESTING EQUIPMENT. The internal resistance of most VOM meters will cause a loading effect on the circuit.

In most cases where a VOM is applied to a solid-state circuit, you may not read anything. For voltage and resistance measurements in these circuits use a FET-VOM, VTVM, or digital meter.

In the solid-state FM receiver you will find the oscilloscope more useful in signal tracing because of the loading effect the signal tracer probe will have on the circuit. Make certain your oscilloscopes have as many as three or four different probe types that are recommended for certain types of circuits.

When troubleshooting the FM receiver, isolate the defective stage either by signal tracing or by signal injection. Then you will have only a few parts to check, making service easier and quicker.

10

Introduction to Semiconductors

The semiconductor is defined as a material whose resistivity is between that of an insulator and a conductor. The resistivity is often changed by light, heat, an electric field, or a magnetic field. Current flow is often achieved by transfer of positive holes as well as by movement of electrons.

The semiconductor may also be defined as a substance whose conductivity is improved by the addition of certain other substances or by the application of heat, light, or voltage.

Summarizing these statements: a semiconductor is the combining of two elements in the proper proportion so that their resistivity is between an insulator and a conductor. Positive holes, as well as movement of electrons within this substance, are achieved by applying an external force of light, heat, electric, or magnetic fields.

Two words used in defining the semiconductor are defined here:

Resistivity: The resistance in ohms that a unit volume of a material offers to the flow of current. Resistivity is the reciprocal of the conductivity of a material.

Conductivity: The ability of a material to conduct, as measured by the current per unit of applied voltage. It is the reciprocal of resistivity.

Electronics usually refers to the semiconductor as solid state. This implies the matter used is a solid, in contrast to liquid or gaseous matter, and conduction of current occurs within the solid material. But this solid-state device is only partially solid; in the "miniature world" of the atom there is only space.

There are many semiconductors. The most popular now are germanium and silicon, which in their natural states are hard, brittle crystals.

An understanding of the molecular or atomic structure is not essential to the construction of a transistor circuit. But to the serious-minded electronic technician with intent to design or troubleshoot solid-state equipment, this knowledge will give him a better understanding of special circuits and solid-state terminology.

To make your understanding easier, we will use a different approach to this somewhat difficult-to-understand technology. We will take a tour inside an element. I will be your guide and explain the events as they appear. We will take this trip via an imaginary magnifying glass that is capable of enlarging the element's interior to such a proportion that it will seem as though you are standing right inside it.

Imagine now that you are looking through a magnifying glass that is focused on an element of germanium. As you adjust the glass, it appears that you are standing at the entrance of an enormous opening. Inside this opening you see what appear to be sphere-shaped objects, all the same size, stacked on top of one another. These are the atoms inside the element, and they align themselves into many different structures. (The word used to denote such patterns of atoms is *lattice*.) Adjust the glass now so that you can look inside one of these round spheres or atoms. Notice, as you adjust the glass, that the atom becomes larger, and that it is not firm and smooth but has a shadowy, hazy look. As the inside becomes visible you are amazed, for it looks like an enormous amphitheather. But instead of the usual seats, you see in the distance what appears to be in the center a dense heavy region. This is the nucleus, and around this center region are small specks that are spinning like a top at very high speed. You notice that there are more than one, and they seem to appear in circles around this nucleus. Notice also that there seems to be more than one of these circles around the nucleus: these are the *electrons*. There seems to be something different about each of these electrons for they are not all spinning in the same direction. Some are spinning from left to right; others are spinning from right to left; all are spinning at a very high speed.

Refer to Figure 10-1. This illustrates the arrangement of the electrons around the nucleus of an atom of germanium. While you are looking at the center nucleus, I might explain to you that it contains two fundamental particles, which are referred to as *mass* and *charge*. Mass means the quantity of the matter in the object, or it is related to

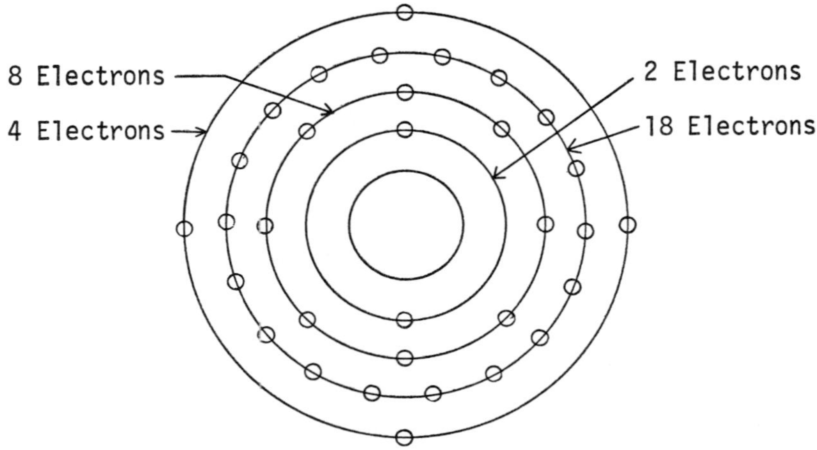

FIGURE 10-1: Arrangement of electrons in germanium atom.

the gravitational attraction. Charge means the quantity of electricity in the particle. There are two kinds of charges, positive (+) and negative (−); so in the nucleus we have both positive and negative charges. The two types of particles in the nucleus are the proton with a charge of +1 and the neutron that has no charge.

Now, refocus your imaginary magnifying glass until you are on the outer edge of that vast space which resembles a large amphitheater. We are now closest to the outer ring or orbit of electrons that are traveling around this center nucleus much as shown in Figure 10-1. Now that we are closer to these electrons, we can see that they are spinning at high velocity. As they pass by like the objects on a merry-go-round, we can see that some are spinning in one direction and others in the opposite direction. But we also notice something else; they do not exhibit a uniform rotating motion. Each one is tilted at a different angle, and some are going through a tumbling motion, but all seem to be rotating at the same speed. As the electron spins, there is a magnetic movement about it; therefore, the electron in motion is an electric current. The electrons orbit around the nucleus, therefore there is a magnetic field about this orbital path that simulates that of a magnet.

Refer to Figure 10-2. This resembles the magnetic path around each orbit circling the nucleus. Looking inside this atom, you cannot see these magnetic fields for the same reason that you cannot see the magnetic fields about a permanent magnet. The lines of magnetic force

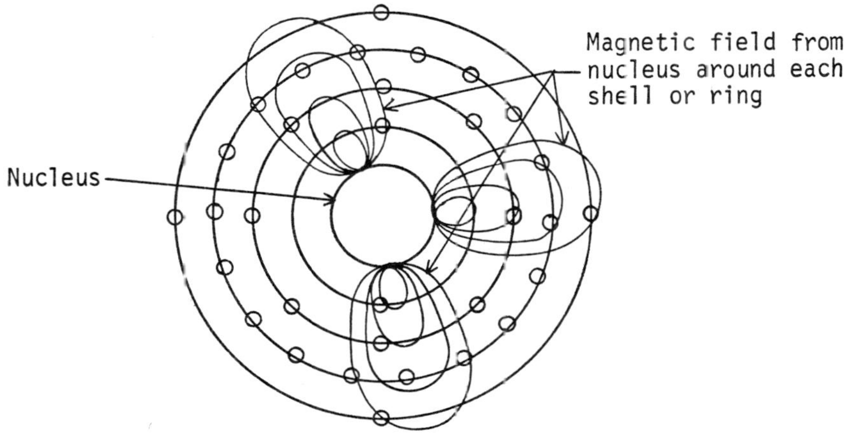

FIGURE 10-2: Magnetic lines of force; each orbit around an electron.

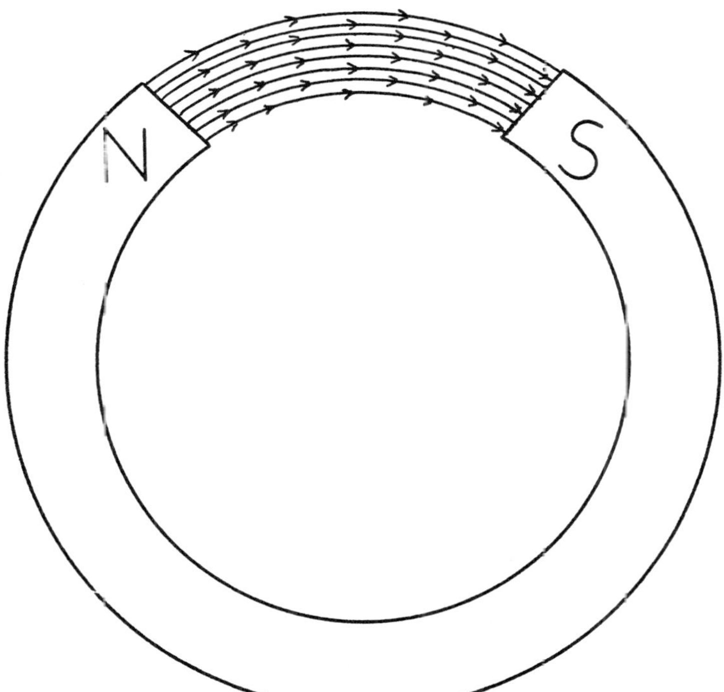

FIGURE 10-3: Magnetic lines of force about a magnet: how they compare to those from nucleus to each shell or ring of the atom.

about a horseshoe magnet are shown in Figure 10-3; we cannot see them but we can measure their effects.

If the magnetic fields about the spinning electrons increase or decrease, they will result in electrostatic fields which produce other magnetic fields. It is thought that in most atoms the greater part of the magnetic effects is cancelled out.

As these electrons spin and travel around freely in the atom, those that are in the last ring away from the nucleus are called the outer valance. These electrons are the only ones in the atom free to leave and travel outside the atom, and all of these electrons you are viewing are negative. Take your eyes from the magnifying glass for a moment and refer to Figure 10.4. Your understanding of the following explanation is probably the most difficult phase in the study of semiconductors, yet it is the most important.

When an electron leaves the outer valence of the atom, the atom is said to have a positive charge. The first illustration in Figure 10-4

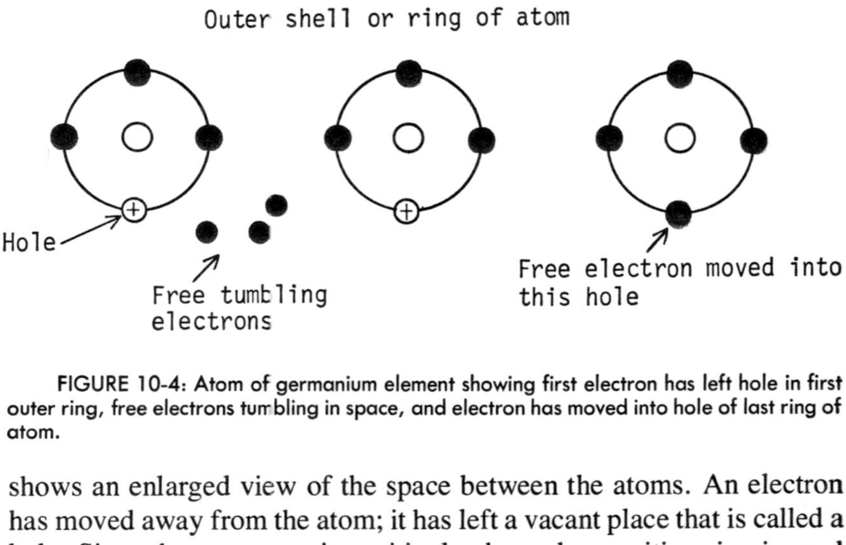

FIGURE 10-4: Atom of germanium element showing first electron has left hole in first outer ring, free electrons tumbling in space, and electron has moved into hole of last ring of atom.

shows an enlarged view of the space between the atoms. An electron has moved away from the atom; it has left a vacant place that is called a hole. Since the atom now is positively charged, a positive sign is used on the hole. This is shown in the second illustration. Since unlike charges attract, the third illustration shows that another atom with a positive hole has attracted this electron.

If we apply a battery to the element we have been viewing, there will be a potential difference between the two ends. (See Figure 10-5.)

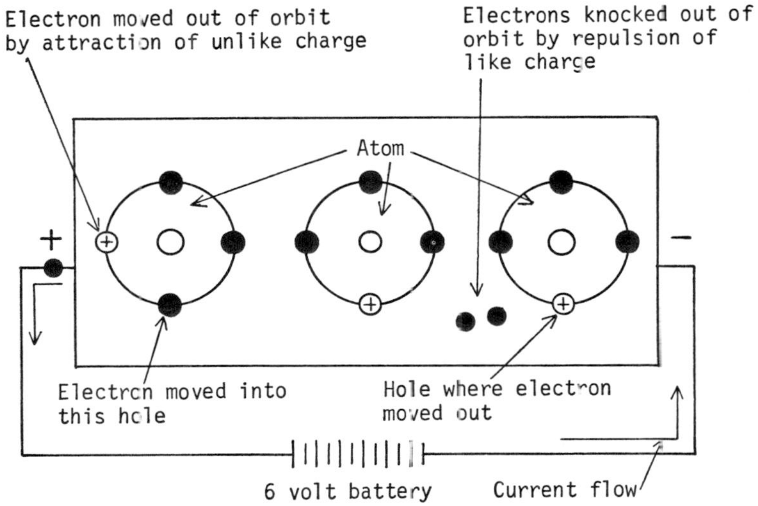

FIGURE 10-5: Germanium crystal chip showing battery connected as an external force to cause electron movement.

This will cause an electric field, and energy will be built up in the form of heat. As shown in the illustration, this energy will force the electrons to leave the atom, and the atom is now positive, as shown in the first illustration to the right at the negative terminal of the battery. Remember that this is called a hole; thus a positive sign is used to indicate its polarity. An electron has moved out of the second atom, leaving a hole, and another electron moves into this hole. The electron that left this hole has already entered the hole of the next atom; this electron was attracted by the force of the positive field at this end of the element.

Refer to Figure 10-6. This is a step farther in the process we are discussing. We have spaced the atoms apart to better explain this movement. Starting from the left hand side, you will notice an electron has moved out of the atom, leaving a hole. This electron is attracted to the positive hole in the next atom, which was left by an electron leaving the second atom.

Notice the arrow from the second and fourth atoms, indicating the electrons are moving from left to right. As they do, they are leaving holes behind. The arrows from the fifth and third indicate the holes seem to be moving from right to left. The electrons are the only

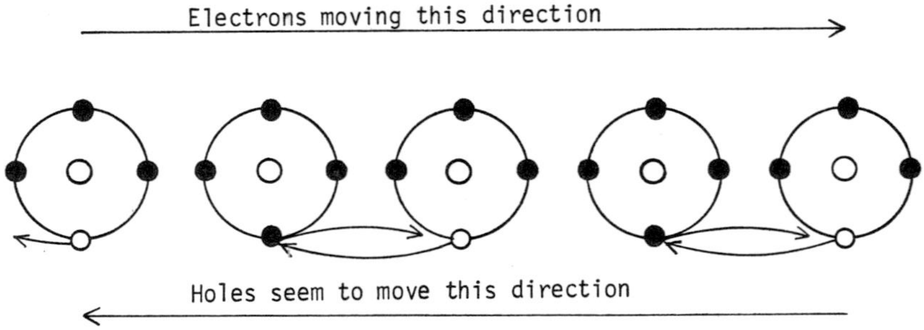

FIGURE 10-6: Illustration of how electrons are moving in one direction and holes seem to move in the opposite direction.

moving particles, but since they are leaving holes behind them, it appears that the holes are moving in the opposite direction. The movement of these electrons is caused by the electric field between a potential difference (positive and negative leads of the battery) which is an energy in the form of heat.

The long arrow at the top of the illustration indicates the direction of the movement of electrons which is current flow. At the bottom of the illustration there is another long arrow, indicating the direction in which the holes seem to be moving. The arrow at the top is the *electron flow* and the arrow at the bottom is *hole conduction*.

Keep this one thought in mind while studying the theory of semiconductors: there is *only one moving particle in the semiconductor—the electron*, which possesses a negative charge.

In the study of semiconductors the hole is just as important as the electron, but what is a hole? The Electronic Dictionary describes it as mobile vacancy in the electronic valence structure of a semiconductor. It is equivalent to a positive charge.

Then—*a hole is a mobile vacancy with the equivalency to a positive charge*.

SEMICONDUCTOR MATERIALS

P- Material:

Germanium and silicon are two elements in the semiconductor class. These elements alone are very weak conductors; that is, their outer valence is very closely bound, and it takes a greater amount of energy to remove the electrons from the atom. To cause them to conduct more readily, they are processed by a method called doping.

In this process another element in small quantity is added to the germanium. Gallium, shown in Figure 10-7(a), is one of the elements that may be added. Notice that near the center is shown one atom of gallium with a hole in its outer valence; this is a deficiency of electrons, and the atom is positively charged. Immediately this atom combines with the germanium atom, and the electron from the germanium atom enters the hole of the gallium atom. One of the atoms in the germanium element is positive because it has given up one of its electrons. The germanium element now has a deficiency of electrons and becomes a positive material referred to as a P- material. Current flow within this material will be by hole conduction. Current flow through the circuit will be electron flow.

N-Material:

If a different element, such as arsenic, is added to the germanium element, a different material will be formed.

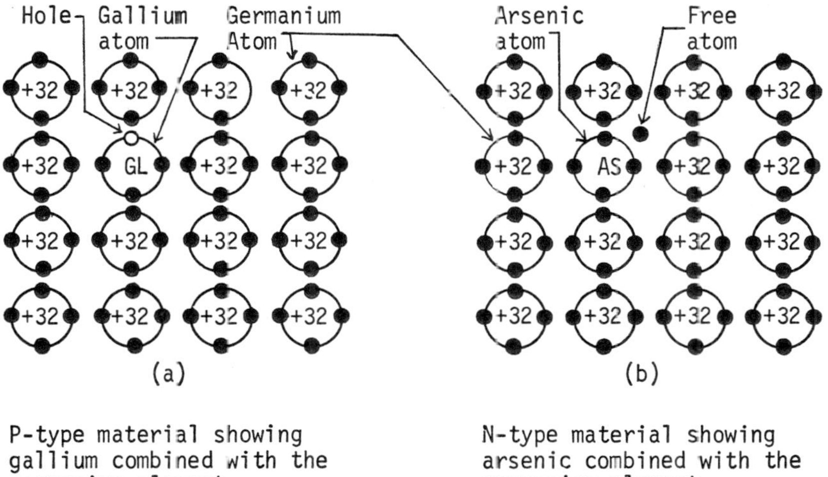

P-type material showing gallium combined with the germanium element

N-type material showing arsenic combined with the germanium element

FIGURE 10-7: P and N type materials.

Refer to Figure 10-7(b), which illustrates the condition that exists when these two elements are combined by the doping process. Arsenic element is shown in (b) as the letters AS, which is the symbol for arsenic. Notice the excess electrons free to wander about within the atom. The material has an excess of electrons; therefore, it is negative and is called N- material. *Current flow within this material will be electron movement. Current flow through the circuit will be electron movement.*

I remind you again of a very important part of the theory that was explained in the first part of this chapter: the electron is the only particle free to move *inside* or *outside* of both materials. It is the movement of the electron in the P- material that makes the holes seem to move; this was called hole conduction. This indicates that the holes only made it easier to conduct current through the material.

P-N Junction and Current Flow

Figure 10-8 shows a block diagram of the P and N type materials. The little circles with a plus sign (+) in the center illustrate the holes, and the circles with the minus sign (−) in the center illustrate the electrons. In this diagram, (a) shows the P- material, which is positive; current through this material will be by *hole conduction*; (b) shows the N- material, which is negative; current flow through this material will be by *electron movement.*

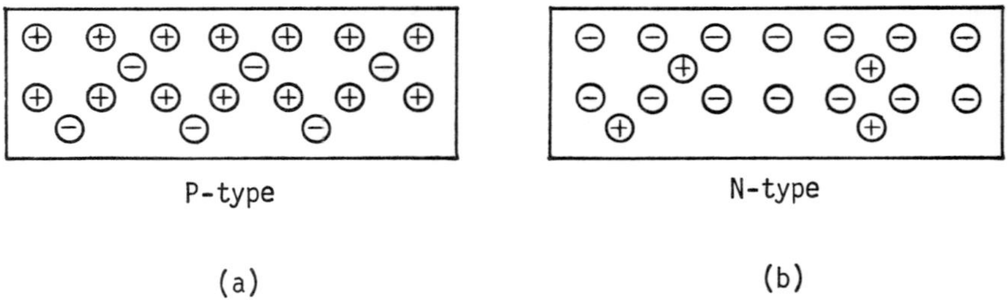

FIGURE 10-8: P and N type materials showing the major current carrier for each material.

Figure 10-9 shows the P and N materials brought together by a method called *fusion.* The two materials are heated until they become a gel; at the correct moment they are removed from the oven and allowed

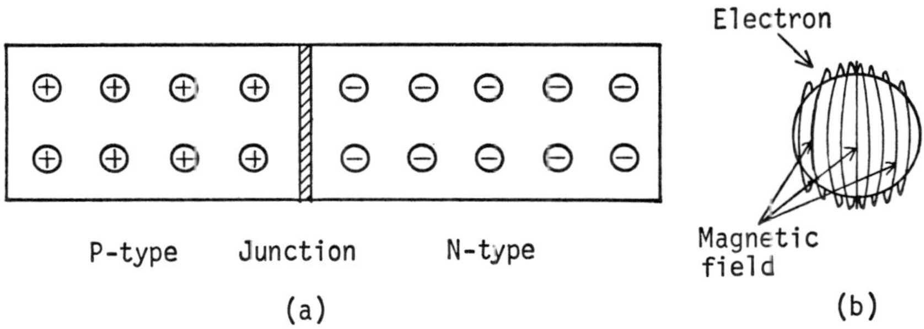

FIGURE 10-9: (a) P-N junction. (b) Magnetic field around the electron.

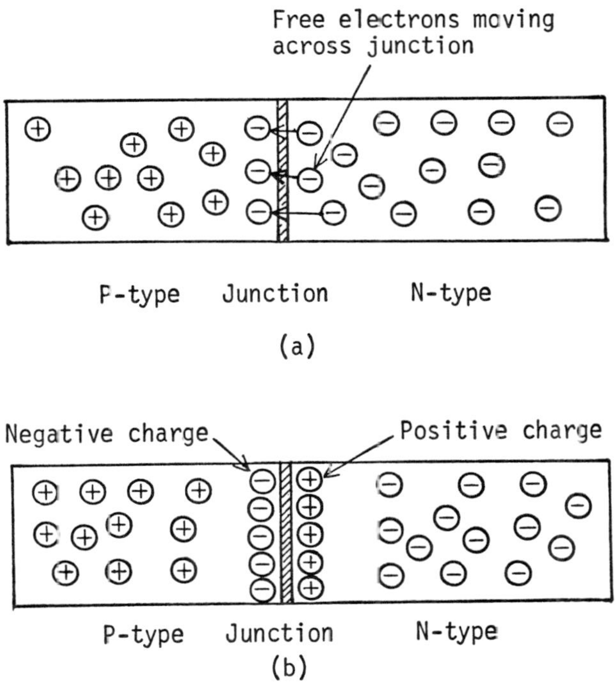

FIGURE 10-10: (a) Electron from N-type material crossing the junction into the P-type material.
(b) Barrier formed at junction by electrons in the P-type material and holes in the N-type material.

to cool. By this method the materials join together, and the point of contact is called the *junction*. The device that has been formed is called the *junction diode*. Its unique property is that it will pass current in one direction only. Its application is rectification and detection.

Figure 10-9(b) illustrates what the magnetic field about the electron might look like; this was discussed earlier in this chapter. When the P and N materials were heated during the process of fusion, a thermal energy (heat) was applied. This caused the electron to leave the atom, the magnetic field to decrease, and the electrostatic field to increase. At this time the electrons in the N material have gained enough energy to cross over the junction and are attracted into the holes in the P material.

When the thermal energy is decreasing (material cooling), the electrostatic field is also decreasing and the electrons are losing their energy. The electrons that left the negative material have left holes near the junction. The electrons that moved across into the P material have caused electrons to be located near the junction in the P material. This action is shown in Figure 10-10 (a) and (b). Illustration (a) shows the electrons moving across the junction; (b) shows how the electrons and holes line up near the junction. In the region between the holes and electrons there are neither holes nor electrons. This area is called the *barrier region*, with electrons lined up on one side and holes on the opposite side. Figure 10-11 shows the dark area as the barrier region, but notice how a small potential is built up, with positive holes on one

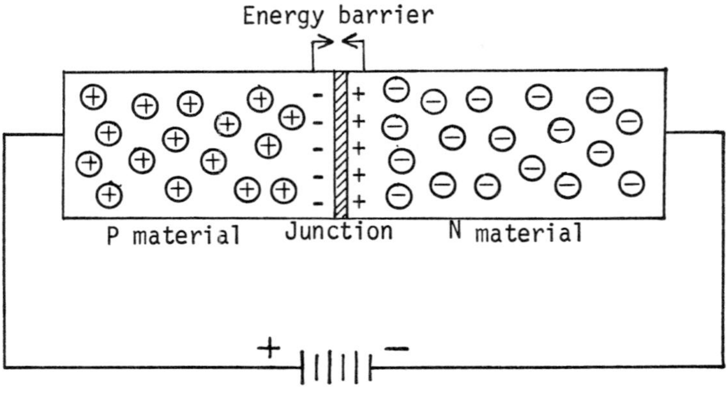

FIGURE 10-11: Forward biased P-N junction diode.

side and electrons on the other. This potential is weak, due to the span across the barrier, but it does form a small battery, which is called a potential hill battery. This is illustrated by the small arrows.

Current Flow

Combining the N and P materials forms a junction diode. Electron movement in a conductor constitutes an electric current, and a potential difference will create an electric field. This electric field receives its energy from a thermal energy in the form of heat. If an electron is placed in an electric field, it will cause the electron to rise to a new energy level and leave the atom.

Refer to Figure 10-11. Notice that a battery has been connected to the diode. This is referred to as forward biasing a junction diode. The positive lead of the battery is connected to the P- material, and the negative lead is connected to the N- material. The polarity of the forward bias will always be the same across the junction; that is, the positive lead of the battery is connected to the positive material, and the negative lead of the battery is connected to the negative material.

Figure 10-12 illustrates what takes place when the forward bias is

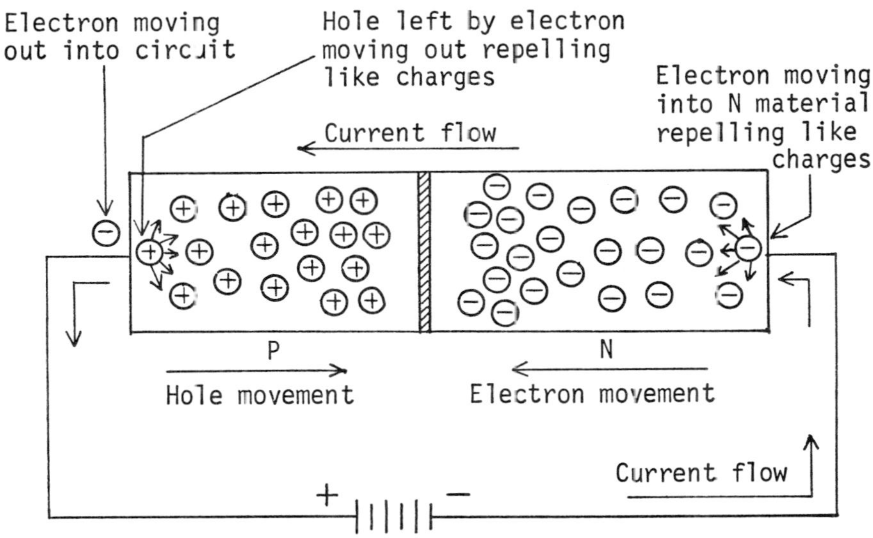

FIGURE 10-12: Current flow when P-N junction is forward biased.

connected. The negative lead of the battery is connected to the N- material, and the positive lead of the battery is connected to the P- material.

A strong electric field is introduced at the negative terminal; this raises the energy level of the electrons in the N- material. The electrons are broken loose from the atoms and are being forced toward the junction. Since the electrons have gained enough energy, they now cross over the barrier region. After crossing over into the P- material, they enter the holes. They continue being forced toward the positive end of the field, the positive terminal of the battery. The positive end of the field immediately attracts the electron outside of the material into the conductor.

Return now to the negative terminal. When the battery was connected, an electron entered the N- material and started a repelling movement, driving the electrons toward the junction. When each electron left an atom, the atom became positively charged. As the next electrons were moving from right to left, it appeared as though the holes were moving from left to right, or in the opposite direction. This is shown by the arrows directly under each material. The arrow under the N- material shows direction of electron movement. The arrow under the P- material shows the direction in which holes appear to be moving. The arrow at the top shows the direction in which the current is flowing. Current flow is in the same direction as electron movement. The arrows along the conductor to the battery show the direction in which current flows outside in the circuit. Current flow inside and outside the material is caused by electron movement.

SEMICONDUCTOR POWER SUPPLIES

So far, only the block diagram has been used to describe current flow through the junction diode. In Figure 10-13(A), (a) is the same block diagram, while (b) shows the symbol for a solid-state diode. The block diagram was just explained, and the symbol is the same. The arrow under the symbol shows the direction of current flow. The vertical straight line in the symbol is marked the cathode; this represents the N- type material, and the arrowhead is marked the anode, which represents the current folow. The diode is called a one way device, for current will pass in one direction only. *The anode must always be positive and the cathode must always be negative before current will pass through it.*

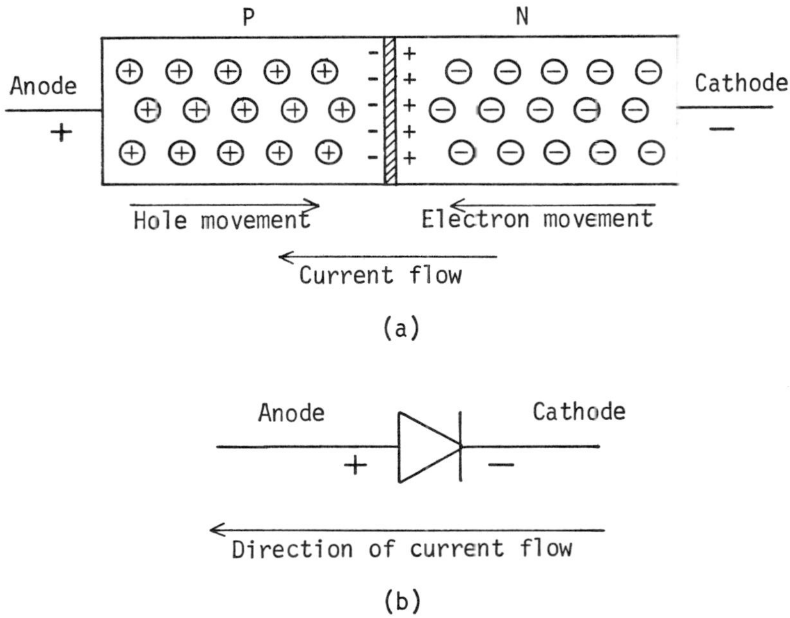

FIGURE 10-13(A): P-N junction diode. (a) Block diagram of diode. (b) Diode symbol.

One of the functions a diode can perform is to rectify. The following explanation describes this function, changing AC to DC.

In Chapter 5 the operation of the power supply was discussed. Here are only a few examples of how the solid-state diode is used and some precautions that are used to protect the diode when it is used in a power supply.

All solid-state diodes have a PIV (peak inverse voltage) rating, and a maximum current rating. In Figure 10-13(B), (a) shows one precaution used to prevent *flashback* in the diode; a capacitor is placed in parallel with each diode.

When the power is applied to the circuit, a part of the voltage is applied to the capacitor, which charges to the full supply voltage. The voltage is also applied to the diodes, and current starts flowing, increasing the voltage across the diode. By the time the voltage has reached the peak of the alternation, the capacitor is fully charged, and the voltage has reached the peak in the diode. This is shown in (b) in Figure 10-13(B). Fully charged, the capacitor has the ability to oppose any change in voltage across it. The capacitor opposes a change in

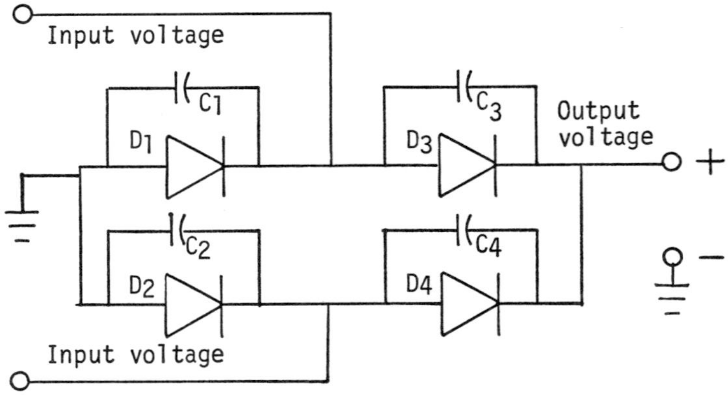

(a)

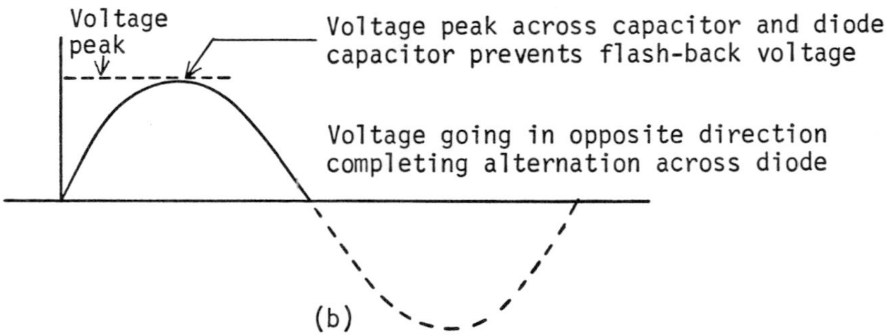

(b)

FIGURE 10-13(B): (a) Capacitors across diodes to prevent flashback voltage. (b) Peak of voltage alternation when capactor opposes a voltage change across diode.

voltage across the diode, and a flashback is prevented. The second half of the voltage alternation continues, only going in the opposite direction and discharging the capacitor.

Figure 10-14 illustrates another protection used in power supplies and especially in the transformerless type supply. R_1 is shown as a current-limiting resistor placed in series with the diode and the supply. When voltage is applied, the resistor receives the surge current, and the current builds up gradually across the diode. The resistor and not the diode takes the surge. This resistor usually has a low resistance value,

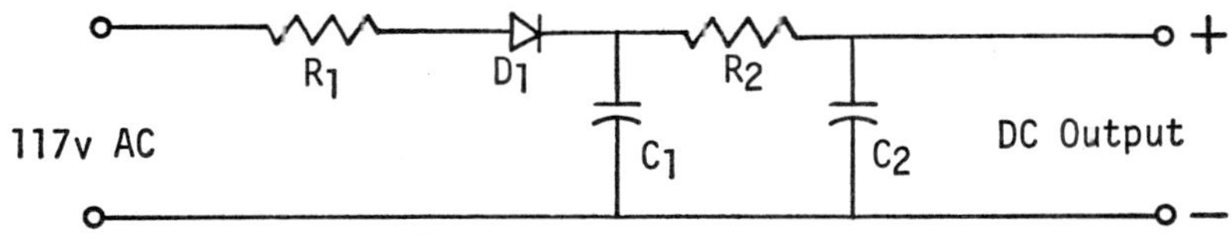

FIGURE 10-14: Half-wave power supply using current limiter R_1 resistor.

and can act as a fuse also. In case of excessive loading or short in the load the diode is protected.

In power supplies used to operate solid-state equipment, it is common practice not to use a ground symbol in the schematic. (In solid-state equipment neither positive or negative is the common tie point or ground.) In the design of the equipment a ground or reference point is established; this is usually common to both negative or positive.

Figure 10-15 illustrates a solid-state full-wave power supply using one large capacitor in the filter circuit. This is common practice in most solid-state supplies. The effectiveness of this type is as good as the convential type of filter network. The capacitor usually has a very large value; many use above 1k mfd. This method of filtering is some times referred to as "brute force" filtering.

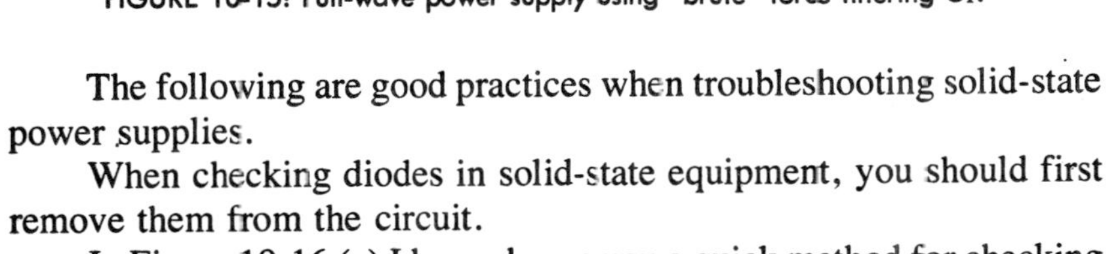

FIGURE 10-15: Full-wave power supply using "brute" force filtering C_1.

The following are good practices when troubleshooting solid-state power supplies.

When checking diodes in solid-state equipment, you should first remove them from the circuit.

In Figure 10-16 (a) I have shown you a quick method for checking a diode after it has been removed from the circuit. Using a VOM connected as shown in the illustration, the meter should read infinity or no reading for a good diode. You are measuring the reverse resistance of the diode. If the meter should read down scale even to two or three

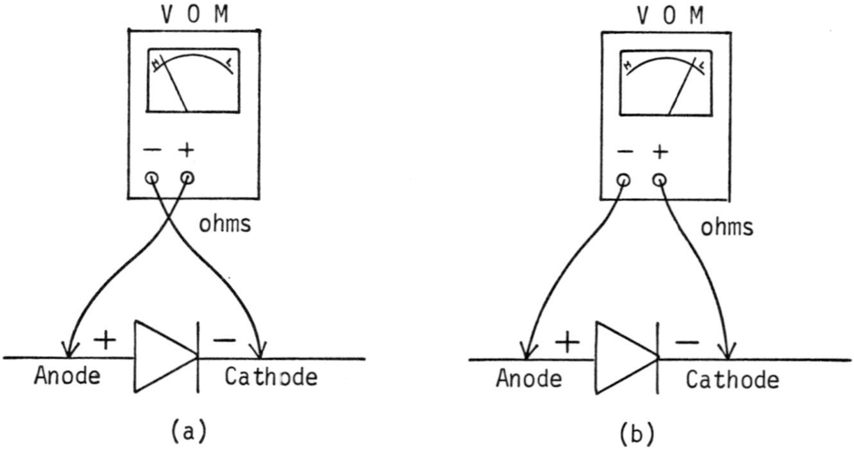

FIGURE 10-16: (a) Checking diode for short or leakage; should show infinity for good diode.
(b) Leads reversed; should show low resistance for good diode.

megohms, do not use the diode; the junction is breaking down and current will flow in the reverse direction.

In (b), with the meter connected as shown, the meter should read a low resistance; different diodes will read different values. If this check should reveal a high value, do not use the diode; output will be low—and may be zero. This check is for the silicon or germanium diodes. When checking the point contact, the little glass bead type, do not hold the meter on it very long. It will be drawing current and will ruin it if it is not already defective.

A quick method for checking a filter capacitor for leakage or short is shown in Figure 10-17. The capacitor should be disconnected from the circuit; discharge it before connecting the meter. Connect the meter as shown in the diagram, negative lead of the meter to the negative lead of the capacitor, and positive to positive. Immediately, when the leads are connected, the meter should kick over to the low range side of the meter. The battery in the meter is charging the capacitor. When the capacitor is charged, the meter needle should start going toward the high end of the scale. The capacitor is discharging and the value of the capacitor will determine how fast the needle will return up scale. If the needle returns up scale above 150k ohms, you may feel sure the capacitor is satisfactory. Any value the meter stops on below 150k ohms indicates the capacitor should be replaced.

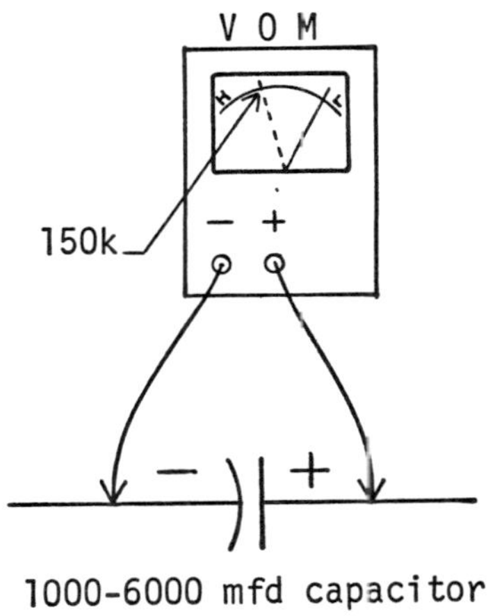

FIGURE 10-17: Method for quick checking filter capacitor used in solid state power supply.

In Figure 10-17, the solid line indicates the meter's first reading. The dotted line indicates the needle has passed above the 150k ohms value; the capacitor is good. If the needle on the meter does not move when you place the lead on the terminals at first, this indicates the capacitor is open. If the needle goes over to the low range and does not return up scale, this indicates the capacitor is shorted.

CIRCUIT CONFIGURATIONS

There are two types of transistors according to their polarity, P-N-P and N-P-N; both are made from the P and N type material. The P- type material current carrier is holes (positive); the N- type material current carrier is electrons (negative).

In Figure 10-18 (a), the top figure illustrates the block diagram of a P-N-P transistor. A small piece of N- type material is placed between two pieces of P- type material. The small circles in the P- type indicate positive holes, while the negative signs in the small circles in the N-type material are symbols for electrons.

The two dark regions on either side of the N- material are the junctions formed when they were made. The process of making them and the explanation given concerning the junction diode are the same for the transistor. I shall not repeat this process again but would sug-

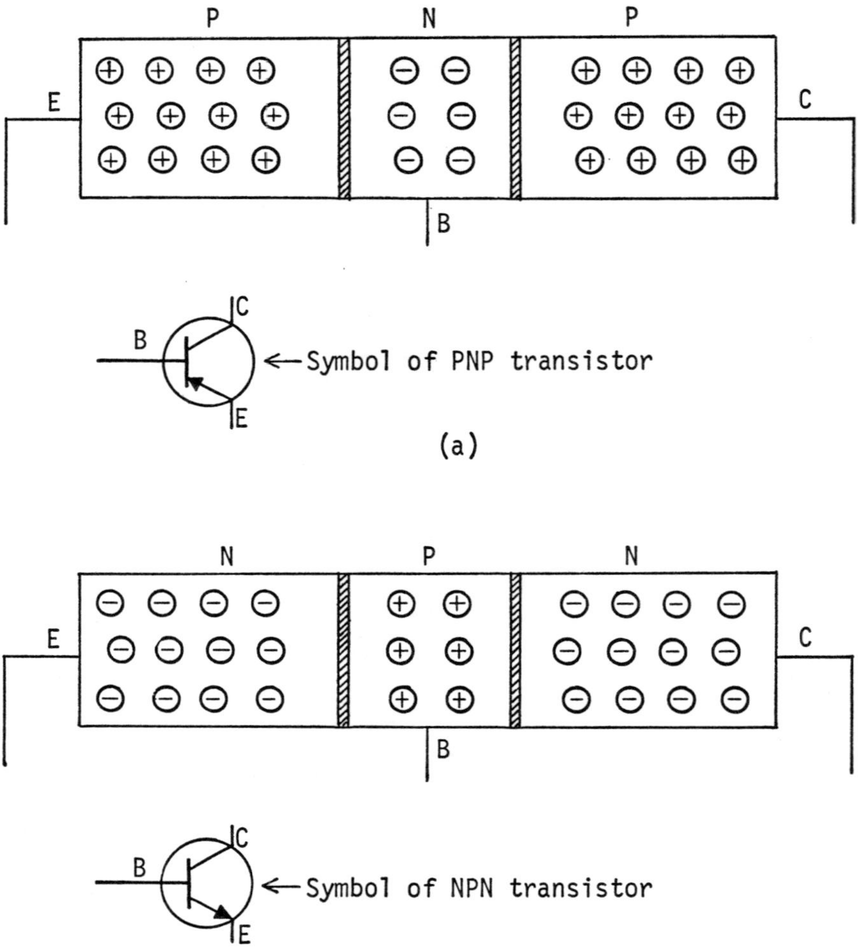

FIGURE 10-18: (a) Shows block diagram and symbol of PNP transistor, showing the element connections.

(b) Shows block diagram and symbol of NPN transistor; also element connections.

gest you return to this explanation on junction diodes if you are in doubt concerning any part of the operation. When you look at this P-N-P block diagram, it does look like two diodes back to back.

The block diagram is marked on the left with the letter E, the center at the bottom is marked with the letter B, and the right side is marked with the letter C.

Directly below the block diagram is the symbol for the P-N-P transistor. Notice that these same letters are marked on the symbol. They represent the elements of the transistor: E—Emitter, B—Base, C—Collector. The emitter is the P- material, the base is the N- material, and the collector is the P- material. In the symbol one of the elements has an arrowhead going into the base element. This indicates in the symbol that this is a P-N-P transistor. Recall that in the symbol for the junction diode the arrowhead was going into the cathode; this indicated the opposite direction for current flow. The same is true with the transistor, for current flow is in the opposite direction of the arrowhead.

This symbol of the arrow pointing into the base of the P-N-P transistor indicates that current flow in this transistor is from the collector to the emitter, and this current flow is by hole conduction. Holes will seem to move from emitter toward the collector, while current flow will be from collector to emitter. This constitutes 97% of the current flow in the transistor. The other 3% flows between the base and the emitter. (More on this current later.)

Refer to (b) in Figure 10-18. This is a block diagram of the N-P-N transistor; a small piece of P- material is placed between two pieces of N- material. The element letters are placed in the same position as the P-N-P transistor. Emitter, Base, and Collector are the three elements for this transistor also. Small circles marked with a − sign indicate electrons and the circles marked with a + sign indicate holes.

Under the diagram is the symbol for the N-P-N transistor. In this symbol the arrow is coming out of the base; this indicates the opposite direction of current flow.

Major current carriers in the N-P-N transistor are electrons. Current flow in this transistor will be in the opposite direction of the arrow. Current flow will be from the emitter to the collector.

COMMON BASE—COMMON EMITTER—COMMON COLLECTOR

There are three types of transistor amplifier circuits, which are illustrated in Figure 10-19. The word *common* is used before each to indicate that element is common to both the input and output.

Illustration (a) is the common base circuit; the base is connected to ground, making it common to both input and output.

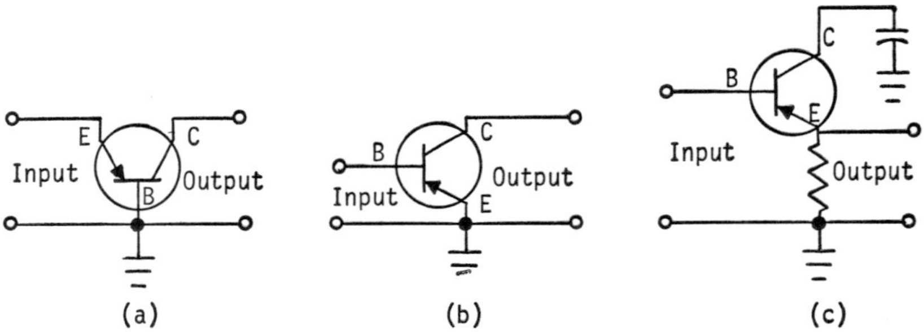

FIGURE 10-19: Transistor amplifier configurations: (a) common base amplifier, (b) common emitter amplifier, (c) common collector amplifier.

Illustration (b) is the common emitter circuit; the emitter is connected to ground, making it common to input and output.

Illustration (c) is the common collector circuit; the collector is connected to ground through the capacitor, therefore it is common to both input and output.

BIASING METHODS AND BIAS STABILITY

Biasing refers to applying a DC voltage to the transistor to cause electron, or hole movement, inside the transistor and current to flow in the complete circuit.

While you were studying the junction diode, I mentioned a very important operation for you to remember. Electrons are the only particles free to move in or out of the atom, or outside the material. As electrons flow in one direction, holes seem to move in the opposite direction. Current flow in a conductor is the movement of electrons.

Figure 10-20 (a) shows a common emitter amplifier using a PNP transistor. The emitter is connected to ground, which is the positive side of the supply.

There are two types of bias with respect to where it is connected to the device. One is referred to as forward bias. *Forward bias is always connected between the base and the emitter.*

The second type of bias is referred to as reverse bias. *Reverse bias is always connected between the base and the collector.* This applies regardless of the type of transistor, PNP or NPN. Also, *the forward bias is always less in value than the reverse bias.*

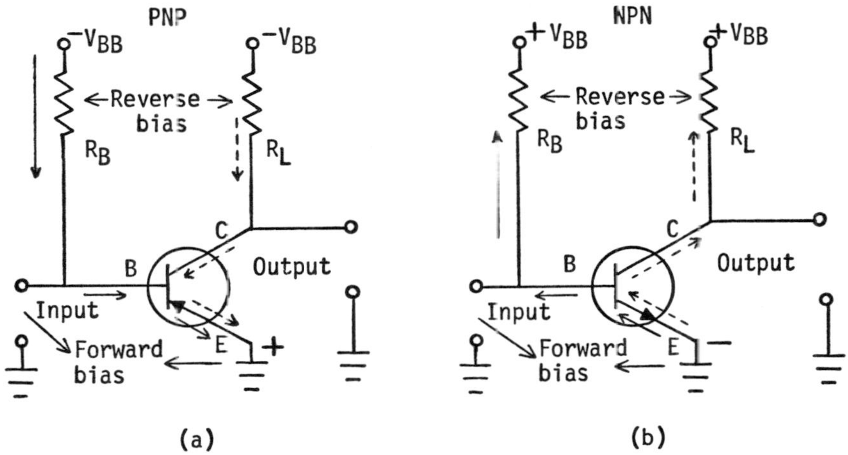

V_{BB} = Supply voltage
R_B = Base resistor
R_L = Collector load resistor
⟶ = Base emitter current
---> = Collector emitter current

FIGURE 10-20: Common emitter biasing showing current flow. (a) Biasing PNP from common power supply. (b) Biasing NPN from common power supply.

Refer now to Figure 10-20: (a) shows arrows at the bottom of the diagram with forward bias marked between them. This indicates that the forward bias is always connected between the base and the emitter. The arrows between the base and collector, indicating reverse bias, are always connected between the collector and the base. Note that both biases are connected to the base, which then controls the current flow through the transistor.

This transistor is a PNP type. Forward bias polarity is connected, positive to the positive material (emitter); and reverse bias is connected, negative to the positive material (collector).

Forward bias is developed across Rb, which is the base resistor. The base resistor is connected to the same polarity as the collector. This establishes forward bias since the emitter is connected to the positive side of the supply. When the supply is connected, current will flow down through Rb, developing a voltage drop across this resistor. The difference between this drop and the supply is the base voltage or forward bias. The current flows through the base into the emitter region and to the opposite terminal of the supply. This is indicated by the solid arrow. Note that current is flowing in the opposite direction to the arrowhead on the emitter. This is the base current and is approximately 3% of the total current flowing through the transistor. This current controls the 97% which is the major current flowing from collector to emitter.

R_1 is the collector resistor. When voltage is applied to the circuit, current will flow down through this resistor, causing a voltage drop across it. This voltage drop subtracted from the supply determines the reverse bias. The current flows through the collector region across the base and out through the emitter to the opposite terminal of the supply. This is a PNP transistor, and *major current flow is always from the collector to the emitter*. The small current in the base is controlling a large amount in the collector-emitter region. In this diagram note that all of the current through this transistor is passing through the emitter.

Refer to diagram (b) in Figure 10-20. This is an NPN transistor, the arrow coming out of the base. (The polarity of the NPN is reversed from that of the PNP transistor.) The forward and reverse bias is between the same elements as the PNP; the purpose of the two resistors is the same. Base current flow is from the emitter to the base; this is forward bias. The solid arrows indicate the direction of the base-emitter current, while the dotted line arrows show the direction of the collector-emitter current.

Some things to remember concerning bias are as follows:

1. Forward bias is always applied between the base and the emitter.
2. Reverse bias is always applied between the collector and base.
3. The PNP transistor, the base-emitter current, always flows from the base to the emitter. Collector-emitter current always flows from the collector to the emitter.
4. The NPN transistor, base-emitter current, always flows from the emitter to the base, and collector-emitter current always flows from the emitter to the collector.

BIAS STABILIZATION

I mentioned earlier in this chapter that temperature is one of the hazards of the transistor, or of any solid-state device.

During normal operation of the transistor, if temperature in the collector should rise, the current will increase also. If this were al-

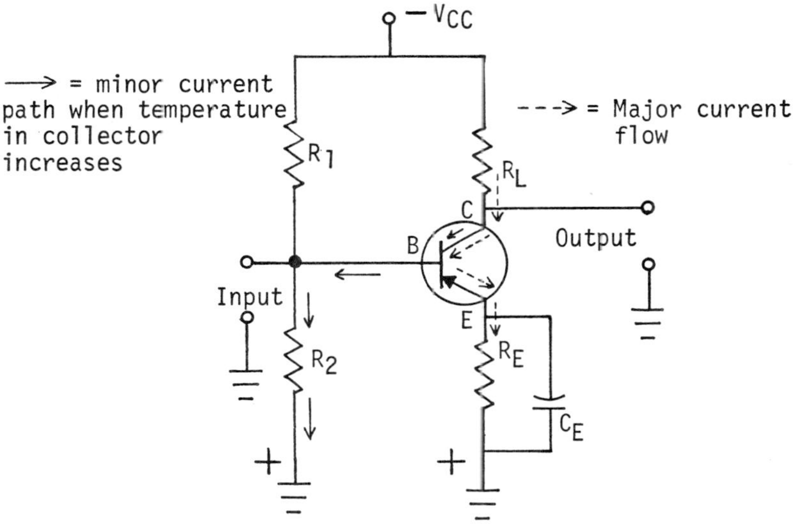

FIGURE 10-21: Current feedback for bias stabilization.

lowed to increase, the base would overheat and damage the transistor. While the current was increasing, current would start flowing in the reverse direction from the collector to the base. This current is called the leakage current. Heat sinks are one way of combating the temperature effects on transistors. This is usually a piece of corrugated metal attached to the body of the transistor, which aids in dissipating the heat.

Power transistors that dissipate more than two watts of power are mounted with a piece of mica between the body and the chassis. The chassis then acts as the heat sink.

Current feedback is another method used to combat temperature variations in the collector region, referred to as bias stabilization. Figure 10-21 shows a resistor placed in the emitter lead to ground. As the current in the collector region increases, due to a rise in temperature, a greater voltage drop will develop across this resistor. This increased voltage will cause the base-emitter junction to become reversed biased. This will cut off the transistor, and the excess electrons collected in the collector-base region will be forced across this junction. R_2 forms a path for these electrons to pass back to the opposite side of the supply.

PRACTICAL TRANSISTOR AMPLIFIER CIRCUITS

There are many different types of transistor amplifier circuits in basic form, just as there are in vacuum tubes. Most of the basic circuits are not practical; that is, their usefulness is limited due to power limitations and capacity to perform one function only. But by combining two or more basic stages, a more complex circuit may be designed—capable of greater power, performing more than one function, etc. To describe the many possibilities in circuit design is beyond the scope of this book. I have selected two or three to show you examples of the not-so-common circuits.

An example of complex circuit designing from the basic circuits might be to take the circuit shown in Figure 10-20 (a) and capacitive couple it to the one shown in Figure 10-21. You would have a two-stage audio amplifier capable of being used to play a tape recorder. The amount of volume would depend upon the type of transistor used and the supply voltage available. But from two basic circuits you would have a two-stage audio amplifier that might perform more than one function. For example, this two-stage amplifier could also double as a

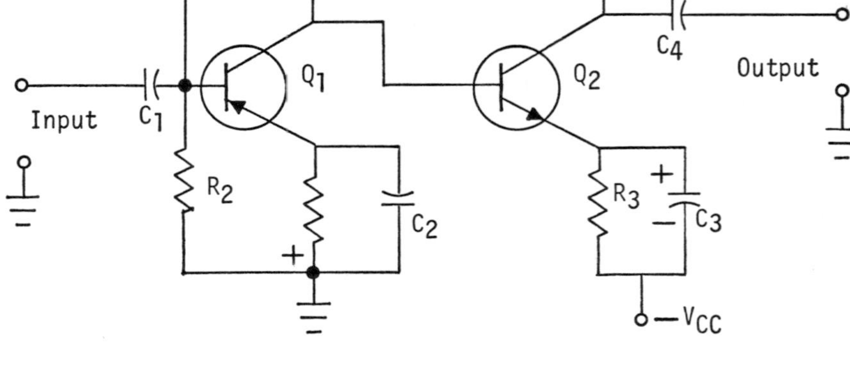

FIGURE 10-22: PNP-NPN direct coupled amplifier circuit.

signal tracer, used to trace an audio signal through other electronic devices. By using an RF probe at the input, it could be used for tracing an RF signal through a radio, or through other devices using an RF signal.

Figure 10-22 shows a type of amplifier circuit using direct coupling. This is not an unusual circuit for solid-state, but it is different from those you have seen in vacuum tube type amplifier circuits. This circuit uses two types of transistor, an NPN and a PNP, with no coupling component between them. This type of coupling is a form of complementary circuit, where the reverse bias of the one stage becomes the forward bias for the next.

Figure 10-23 shows two types of special circuits. Diagram (a) shows the emitter follower circuit, which is used for impedence matching, or as a coupling device between two different stages. When the impedence varies greatly, this amplifier can be used to match two stages. One such example is the RF amplifier stage in the modern television receiver, to match the low impedence of the feed line to the high impedence of the converter stage.

Diagram (b) shows a phase splitter, so called because the two outputs are 180° out of phase with one another. Between the base and emitter there is no amplification, so that the output signal from the

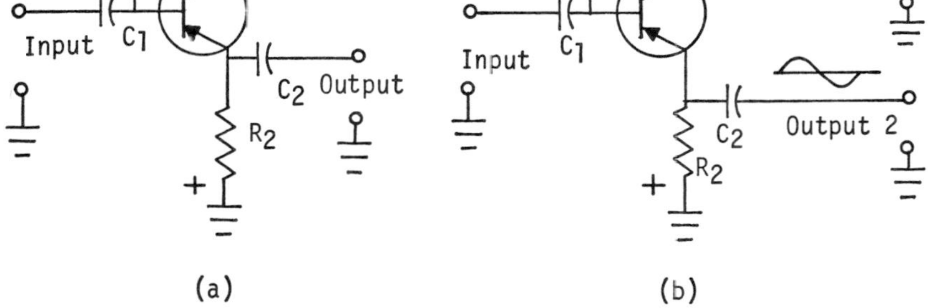

FIGURE 10-23: Two types of amplifier circuits for special use: (a) emitter follower amplifier, (b) phase-splitter amplifier.

emitter is the same as the input signal. But this is a common emitter circuit, and amplification takes place between the base and collector; therefore, the output is 180° out of phase with the input. The two output signals are 180° out of phase with one another. It can be used to drive a push-pull output stage in place of using a transformer, where the input signals to the output signals must be out of phase with one another. It could be used as part of the sync separator circuit in a television receiver, to separate the vertical and horizontal signals from the composite signal.

This circuit could be used in a logic circuit, where pulses of two different polarities are needed from one single input. A positive pulse at the input would give a positive and negative pulse at the output.

TESTING TRANSISTOR CIRCUITS

Diodes, transistors, SCR (silicon control rectifier), IC chips, etc., are not fragile devices (like vacuum tubes). In their physical structure they can withstand great pressures and physical abuse, but they are very fragile as to heat or proper value and polarity bias.

Until you have gained some experience, use caution when working with solid-state devices. The following are some good practices I urge you to follow:

Handling: Many transistors come wrapped in a tinfoil container; the instructions say AVOID TOUCHING WITH YOUR HANDS WHEN INSTALLING. You should follow these instructions to obtain the best performance from the device. Avoid bending the leads of the transistor back and forth. They may break off at the top of the case.

When installing a transistor, it is good practice to leave as much lead as possible, because the leads will act as a heat sink.

When removing transistors from a PC board, never pull on the leads with much pressure, for you can pull the leads out of the case.

When installing a transistor or IC chip into a socket, make certain that all of the pins go into the socket straight. One could bend over, and you might not notice until it would not operate. If you bend a lead or pin, use a pair of small needle-nosed pliers or a pair of tweezers to straighten.

Never expose a FET to a strong magnetic field, because this may cause damage. If you have, or are using, a FET-VOM, never place the leads close to the high voltage in a TV receiver. The probe will pick up the magnetic field and apply it to the transistor and ruin it.

GOOD PRACTICE WHEN USING TOOLS AND TEST EQUIPMENT

Use only a soldering iron that is rated no more than 40 watts. Use a pencil iron that has different types of replacement tips: pointed, chisel, etc.

Use only low-temperature rosin-core solder.

When soldering, apply only the amount of heat necessary for the solder to flow evenly.

Avoid messy solder pile-ups or "globs." This can cause a "cold solder joint," and trouble. Make certain the joint is electrically as well as mechanically secure.

When soldering a printed circuit board, the circuits are usually close together. Avoid holding your iron too long on one spot and use the least amount of solder possible. This is to avoid having the solder run and maybe join two or three circuits together.

When soldering a transistor into a circuit, grasp the lead with a pair of long-nosed pliers between the body and the end of the lead. The pliers will act as a heat sink and avoid damage to the transistor. Avoid holding the iron tip on the transistor lead if possible, and then for only a very short time.

Always make certain of the values needed for the circuit and the polarity of the circuit before applying power.

Never pull transistors from the socket while the power is still applied to the circuit. This can cause a current surge through the device and ruin it.

When using an RF and AF signal generator for signal injection or for alignment, turn the gain on the generator off; then after it has been connected to the equipment, turn the gain up only enough to give satisfactory results. Too much signal will cause the base current to increase beyond the limits of the transistor and ruin it.

Avoid connecting the generator probe directly to the transistor. Holding it close to the circuit will sometimes induce enough signal into the circuit for satisfactory results. Clipping the leads close to the coil will sometimes induce enough signal into the circuit. Connecting a small value capacitor to the end of the probe and then touching the capacitor to the circuit performs very well.

SOME PRACTICAL SYMPTOMS

Use the oscilloscope to measure the input and output signals. Low gain in any stage is almost a certainty of a leakage in one of the junctions of the transistors. Forward and reverse bias are developed across a resistor in series with one of the transistor elements. Measuring the voltage drop across these resistors can tell you what is happening in this stage.

First you must know the type of equipment you are testing. Some circuits are operated with one or more transistors in the cut-off state. For example, if the full supply voltage is being dropped across the collector load resistor, the current is maximum, and the collector is near zero potential. The transistor is in the cut-off state. Or if the full supply voltage is being dropped across the collector load resistor, the current is maximum, and the collector is near zero potential. The transistor is in the cut-off state. Check the emitter and base voltage; their voltage should vary considerably from the collector voltage. The voltage variation between the emitter and base should be on the order of 0.1 to 0.3 volts. If the transistor is cut off, these voltages will vary a great deal.

When a transistor is conducting properly, each resistor should show a voltage drop across it. This indicates that current is flowing in the circuit. If no voltage is dropped across more than two of the resistors, the transistor may be open. If the voltage drop across one of these resistors is higher than normal, you may have a leaky transistor. When the emitter and the base have the same voltage on them, the transistor is cut off. Be careful. Do not condemn the transistor. A leaky coupling capacitor can apply a reverse bias to the base that will cut the transistor off also.

When checking a capacitor, at least one end must be disconnected from the circuit. The same applies to a resistor.

There are transistor checkers that will check the transistor in the circuit, but when in doubt, take the transistor out of the circuit and check.

11

Multivibrators: Astable and Unstable Electronic Circuits

MULTIVIBRATORS DEFINED

Multivibrators are defined as two amplifier stages (vacuum tube or transistor), designed to use positive feedback or regeneration between stages.

Unless you have an understanding of the term feedback or regeneration, the above definition does not give you a clear understanding of this circuit.

Refer to Figure 11-1 and follow the step-by-step analysis. If you divide this circuit into parts, it will look somewhat different to you. Part one and part two are separate common emitter amplifiers with their respective forward and reverse bias resistors. Recall in the study of basic transistor amplifiers that the common emitter was the only amplifier in which the input signal was out of phase with the output signal (inverted). Each of these stages will operate as an inverter, and if the collector of stage two is coupled to the input of stage one by C_2 when voltage is applied to the circuit, the output of stage one will be in phase with that of stage two's input. Also, the output of stage two will be in phase with the input of stage one. As long as the circuit is energized, there will be a positive feedback (in phase) between stages or regeneration which will result in an ideal oscillator circuit. Feedback network and frequency will be discussed later in this chapter.

A multivibrator circuit could be described as two amplifier stages cross coupled in such a manner that the outputs of the two stages are in phase with the inputs, which results in sustained oscillation between the two stages. Positive feedback or regeneration could be described as

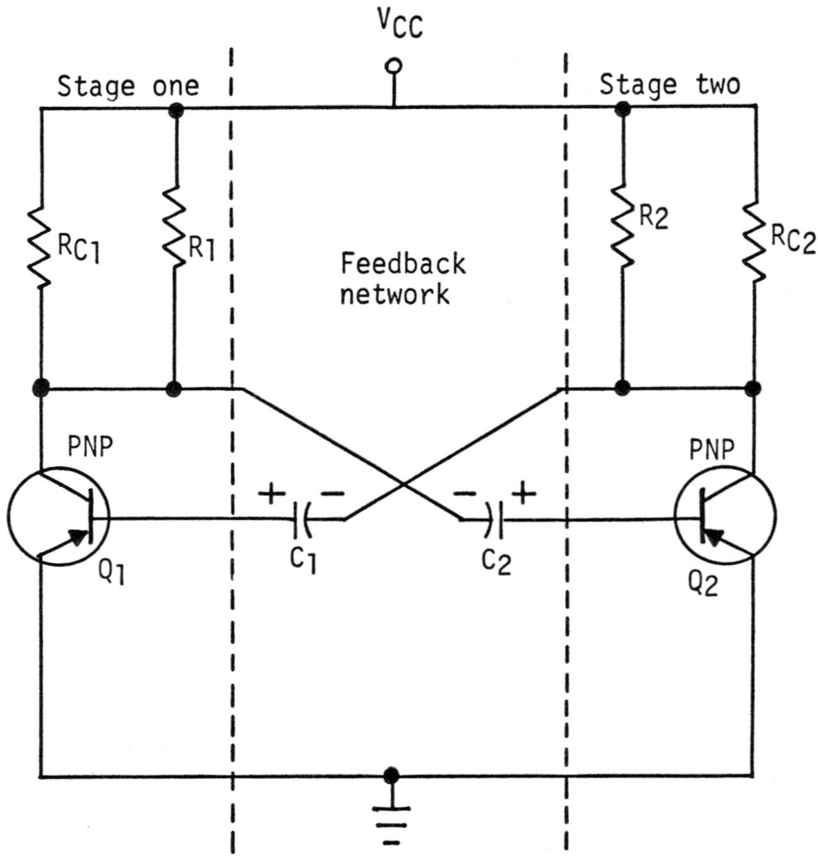

FIGURE 11-1: Multivibrator circuit.

a feedback voltage that will add to the input (in phase) rather than subtract from it.

TYPES OF MULTIVIBRATOR CIRCUITS

There are three general types of multivibrator circuits:

1. astable, shown in Figure 11-2(A).
2. monostable, shown in Figure 11-2(B).
3. bistable, shown in Figure 11-2(C).

These three can be divided into classes which describe the function, the frequency, and type of wave form produced.

Astable, Figure 11-2(A)

This circuit consists of two amplifier stages so designed that when voltage is applied only one stage is conducting at one time; the other is cut off. If the stages are designed to operate alternately (one stage conducting and the other stage cut off), the circuit will act as an oscillator and is called a "*self-excited, free-running*" multivibrator. "Self-excited" means that components in the circuit start the oscillation (because of unbalance in the components or the circuit itself), and that it does not rely upon any outside source to sustain oscillation. "Free-running" is defined as oscillation that will continue as long as the circuit is energized (voltage applied). The astable multivibrator is

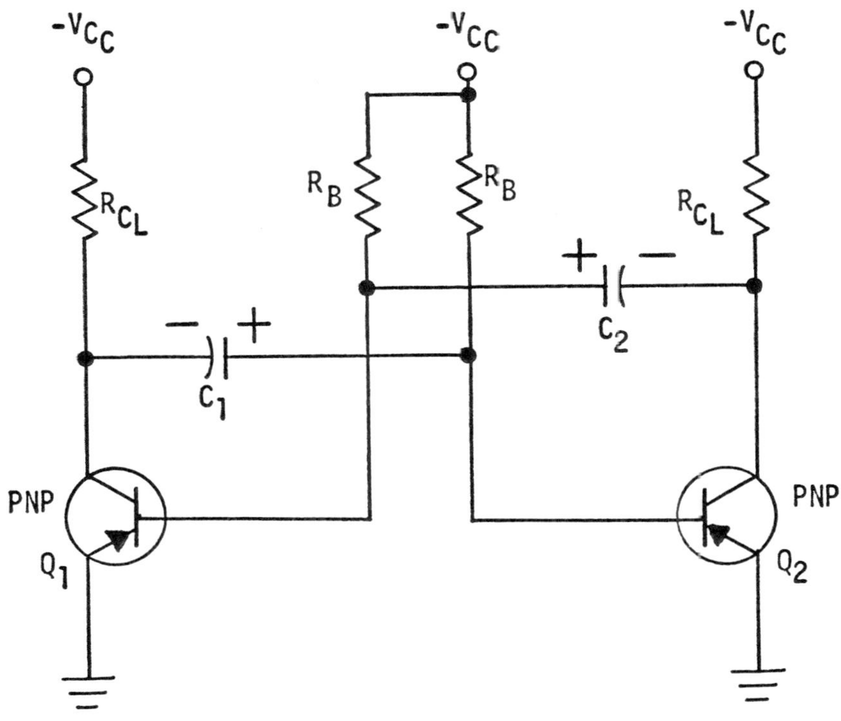

FIGURE 11-2(A): Astable multivibrator.

an oscillator circuit. However, signals can be fed into the input to synchronize the operation and to produce subharmonics. A more detailed description of this circuit will follow in this chapter.

Monostable, Figure 11-2(B)

This is not an oscillator circuit. It operates with one stable state (one side always conducting) and one unstable state (one side off or not conducting). It requires an input pulse to change states which results in the monostable "one-shot" multivibrator.

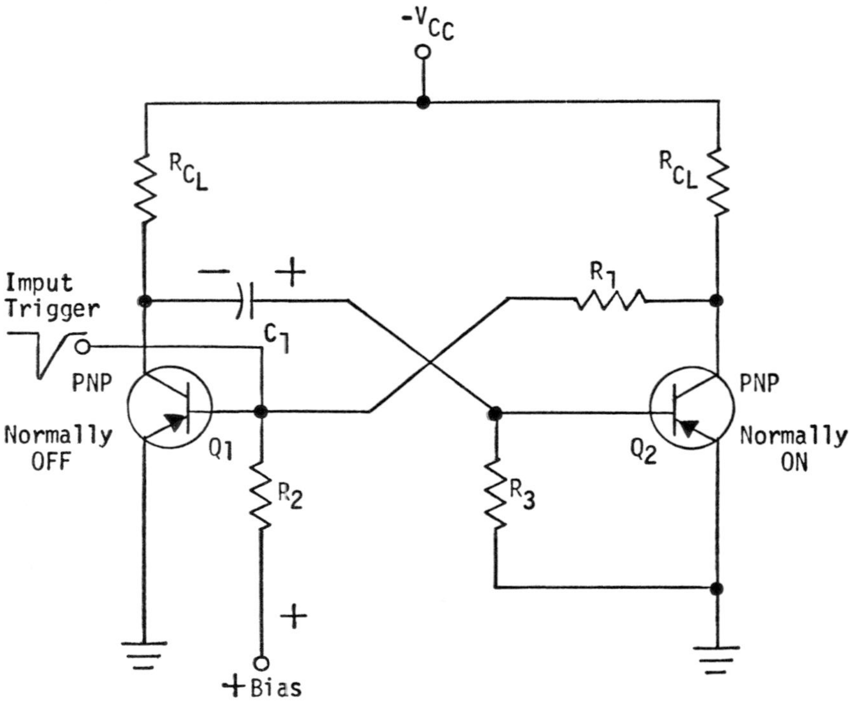

FIGURE 11-2(B): Monostable (one-shot) multivibrator.

Bistable, Figure 11-2(C)

This circuit, commonly called a "flip-flop," is so called because it has two permanent stable states; that is, either of the stages could conduct indefinitely as long as power is applied. The circuit is not an

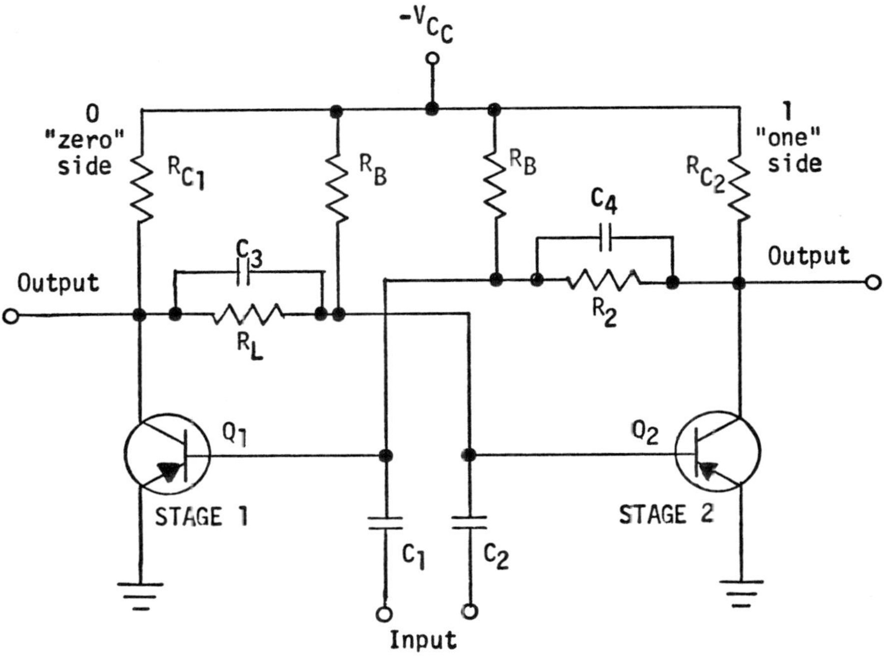

FIGURE 11-2(C): Bistable multivibrator.

oscillator but requires an input trigger pulse to flip from one stable state to the other and a second input pulse to flop it back to its original state. It takes two input pulses to produce a single output. This circuit is commonly used in logic to count binary numbers and to divide frequency, as well as for many other applications.

For an analysis of each of these stages, refer first to Figure 11-2(A). This is the astable or "free-running" multivibrator oscillator circuit.

To start this theory of operation, assume that Q_1 is on and Q_2 is off. If Q_1 is on, it is conducting heavily and its collector is near ground potential, and if Q_2 is off, its collector is equal the supply. Capacitor C_2 charges to almost the supply; the right side of the capacitor is negative. Since Q_1 is on, its base is almost equal the negative supply. With Q_1 saturated, its base is near ground potential; therefore the left side of the capacitor is near ground potential.

Suddenly Q_2 starts to conduct and goes immediately into saturation. But the capacitor is slower in discharging (remember it takes five time constants), and the capacitor continues to discharge. Q_1 has cut off now and Q_2 is conducting heavily and C_1 has started to charge up toward the supply voltage. At this time C_2 has almost completely discharged, but before it can completely discharge Q_1 starts conducting. The base has gone sufficiently negative to turn Q_1 back on. By this time C_1 has almost completely charged up to the supply. The same action takes place now as did before. Q_1 comes back on and Q_2 starts turning off; C_1 starts discharging and C_2 starts charging up again. Now one cycle has been completed.

Figure 11-2(B) shows a step-by-step analysis of a monostable circuit. A stage that is conducting heavily (drawing current) is considered the *on* stage, and the stage that is not conducting (not drawing current) is considered the *off* stage. Assume, then, that Q_2 is the on stage and Q_1 is the off stage. The collector of Q_1 is coupled to the base of Q_2 through C_1, and Q_2's collector is coupled to the base of Q_1 through R_1. Q_1's base is coupled to the high positive bias voltage through R_2, and this, being a PNP transistor, reverse biases the base and keeps the collector current cut off. The base of Q_2 is returned to zero (ground) potential, which causes its collector current to be high. In the absence of an input trigger pulse, the circuit remains in this stable state indefinitely. When a high negative input pulse is applied to the base of Q_1, the normally off stage, the state is temporarily reversed; and the stage Q_1 now becomes the off stage. After a period of time determined by the constants C_1 R_3, the circuit returns to its original state, remaining there until the next incoming pulse arrives. The process has resulted in the name "one-shot." One pulse causes it to change from one state to another but it returns to the original state on its own circuit components.

Figure 11-2(C) illustrates a practical bistable multivibrator circuit or flip-flop. Follow the diagram while the circuit is being analyzed.

Assume that stage two is conducting and stage one is cut off. Q_2 in stage two is a PNP transistor; for this transistor to conduct, the polarity of the base must be negative with respect to a positive emitter. This condition will cause current to pass through the collector load resistor, R_{C2}, and drop most of the supply voltage; the collector will be at a very low potential (low negative voltage). Applying to the base of this stage a positive pulse opposes the forward bias of Q_2, causing the current in the collector to decrease. This decrease in current reduces the voltage drop across the collector load resistor, R_{C2}, and causes the

Q_2 negative collector voltage to rise. This rise in negative voltage is fed to the base of Q_1 by R_2, C_4 feedback network causing it to be forward biased, and Q_1 will start conducting. This regenerative action will continue until Q_2 is cut off and Q_1 is conducting. The input pulse causes a flip from one stage to another. Now stage one is in a stable state (conducting), and stage two is cut off (nonconducting). This condition will remain until the next pulse arrives. If a positive pulse is applied to the stage one, there will be the same condition as before, only stage one will start cutting off because of the reverse bias being applied to the base by the positive pulse. The regenerative action of the feedback network of R_1, C_3 will cause the increasing negative voltage from the collector of Q_1 to be applied to the base of Q_2, starting it to conduct. The regenerative feedback will continue until Q_1 is cut off and Q_2 is conducting. The pulse applied to the input has returned the stage or flopped it back to the original condition and has completed the cycle, causing one output pulse. Two input pulses were required to produce one output pulse. Capacitors C_3 and C_4 were added in parallel with R_1 and R_2 to speed up the flip and flop between stages. The combining of the resistor and the capacitor in the feedback network acts like a differentiating circuit, putting a "spike" in the voltage pulse which causes a more rapid action of flip to flop (or turning on or off of the two stages).

Consider this circuit as a logic circuit, and use it to count in the binary number system which has only two numbers, *zero* and *one*. Assign a *zero* to the on (conducting) stage and a *one* to the off (nonconducting) stage. Call stage one the *zero* side and stage two the *one* side. Follow through Figure 11-2(C) now, and assume the following conditions or states.

Zero State

The flip-flop circuit is considered to be in the zero state when stage one or *zero* is nonconducting (collector voltage is high negative) and producing an output while stage two or *one* is conducting (low negative voltage) and not producing an output.

One State

The flip-flop circuit is considered to be in the one state when the zero side is conducting (collector voltage is negative low) and producing no output while the one stage is nonconducting (collector voltage is high) and producing an output.

The above conditions can be stated in the following table:

Stage or	*State*	*Condition*	*Output*
Zero	Off	Negative collector voltage high	1
One	On	Negative collector voltage low	0
Zero	On	Negative collector voltage low	0
One	Off	Negative collector voltage high	1

LOGICAL SET AND RESET TYPE FLIP-FLOP

The bistable multivibrator can be considered an electronic switch and is capable of storing information.

Figure 11-3 shows a symbol for the RS-type flip-flop. The following information is used to describe this logic symbol.

The letters FF in the block are for flip-flop, and they may be followed by a number according to the order in which it appears in the circuit. The letter S on the left side is for *set*, and R is for *reset*. On the right side, letter A is for one output while $\overline{A}$ (not A) is the other output. Assume the following analysis.

A signal applied to the R input puts the flip-flop in the reset state. $\overline{A}$ output equals one (1), and A equals zero (0). The circuit will remain in this stage until an input signal is applied to the S (set) input. Any signal applied to the R input will not change the state. A signal applied to S will change the state from reset to set. The output now will be A equals one (1), and $\overline{A}$ equals zero (0). The circuit will remain in the set state until a signal is applied to the reset or R input. The switching action of the flip-flop caused by the input pulses between reset and set has caused the circuit to store a 0 in the reset state and a 1 in the set state. This information will remain in either of these states until an input pulse is applied to the opposite state.

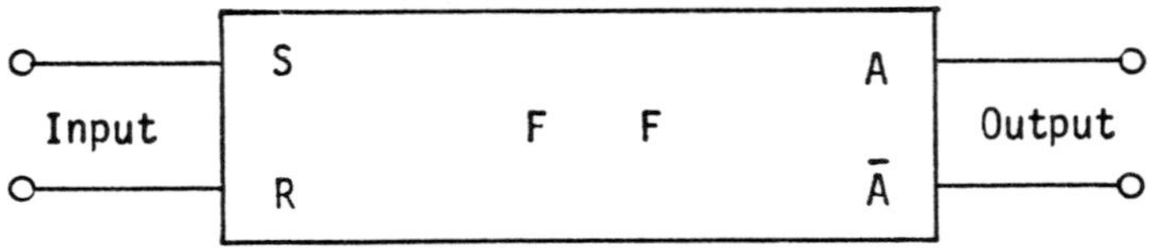

FIGURE 11-3: The top line only is logic symbol for RS type flip-flop.

USING TEST EQUIPMENT AND LIVE CIRCUITS TO PROVE THEORY

The theory of the astable, the monostable, and the bistable multivibrator circuits has been analyzed. Now you will learn how this theory operates by constructing a practical circuit of each of the different circuits. You will use test equipment and other means to see the theory come alive in operation.

Figure 11-4(A) shows a diagram you might construct of a practical astable multivibrator. The value of the capacitors C_1 and C_2 was made large in value to slow down the condition of flip-flop so that it could be viewed on the meter. If a smaller value of capacitor had been used, the action of the meter would have been too fast to follow. You are using the meters to measure the voltage drop across the collector load resistors R_{C1} and R_{C2}. It was stated earlier in this chapter that

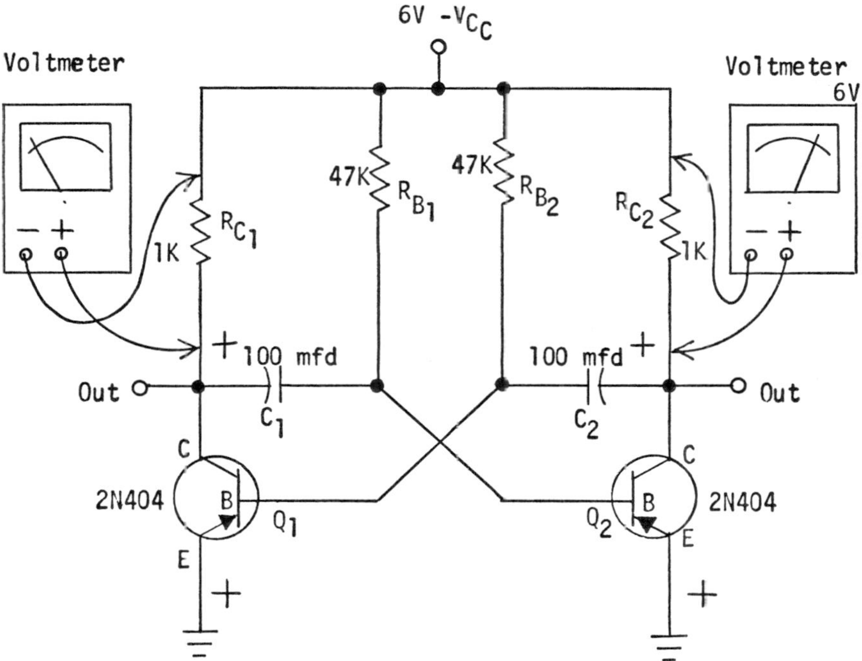

FIGURE 11-4(A): Astable multivibrator showing method of connecting two voltmeters across the load resistors of the two stages to measure the voltage drop across each resistor.

when power was applied to the astable circuit, one state would start conducting first and the other stage would be cut off, and that the condition would reverse periodically, flipping and flopping back and forth, a condition we call oscillation. You also learned that the time between flip and flop or conducting and nonconducting was measured as the frequency of oscillation.

Connect the supply voltage (battery or DC supply) to the circuit. Notice that one of the meters is reading zero; this stage is not conducting. No current is passing through the resistor in the collector circuit; therefore, there is no voltage drop across this resistor. The meter that is reading a voltage shows this stage is conducting and that current is passing down through the resistor into the collector, across the base, into the emitter, and back to the supply. This condition will produce a voltage drop across the resistor which the meter is reading. Notice that periodically the meter will reverse and continue reversing as long as the circuit has power applied. This reversing action is showing you the flip-flop condition that is taking place in the circuit as one stage conducts and the other stage is cut off. In order to show you further proof of this flip-flop action, move your meters as shown in Figure 11-4(B). If you move the positive lead to the collector, the other meter is placed in the same manner on the collector and ground. You are now reading the collector voltage of each stage as it conducts. When one meter is reading zero, it shows that this stage is conducting, and all of the

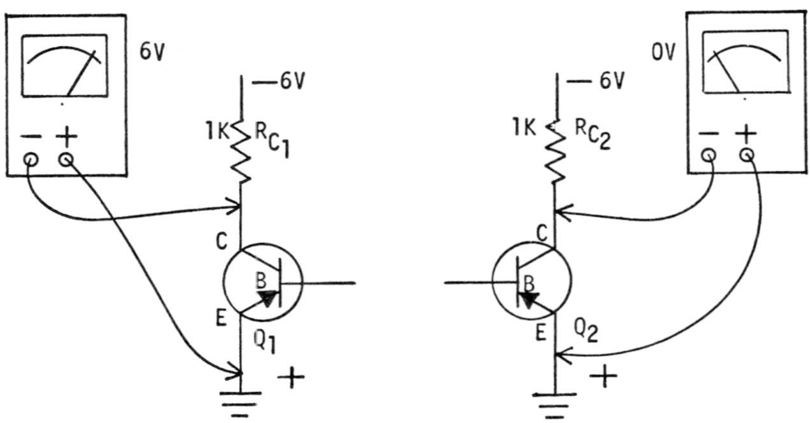

FIGURE 11-4(B): Voltmeters connected to read collector voltage of the two stages.

supply voltage is being dropped across the resistor with no voltage appearing at the collector. With no voltage on the collector, there is no output produced, while the opposite side collector shows a voltage and is producing an output. After a short time (determined by feedback network C_1, R_b), the stage will flip, Q_2 will start conducting, and Q_1 will cut off. The meters will reverse. The one on the left will read voltage, and the one on the right will read zero. The meters in (A) will be reversing in the opposite direction of those in (B). Meters placed at either place will show the action of the conducting and nonconducting stages and will also show the stage that is producing an output pulse.

Another method of viewing the action inside the circuit of the astable multivibrator is shown in Figure 11-4(C). Connect your oscilloscope to this circuit as shown. You should see a near DC pulse as the two stages operate between conducting and nonconducting. The wave form shows the slow time of the RC timing circuit. As the capacitor charges and discharges, which is the frequency of the circuit, notice that the wave from appearing on the oscilloscope may not be a perfect square wave. The reason is that both networks are not symmetrical (are not exactly the same value), or the characteristics of the two transistors may not be the same.

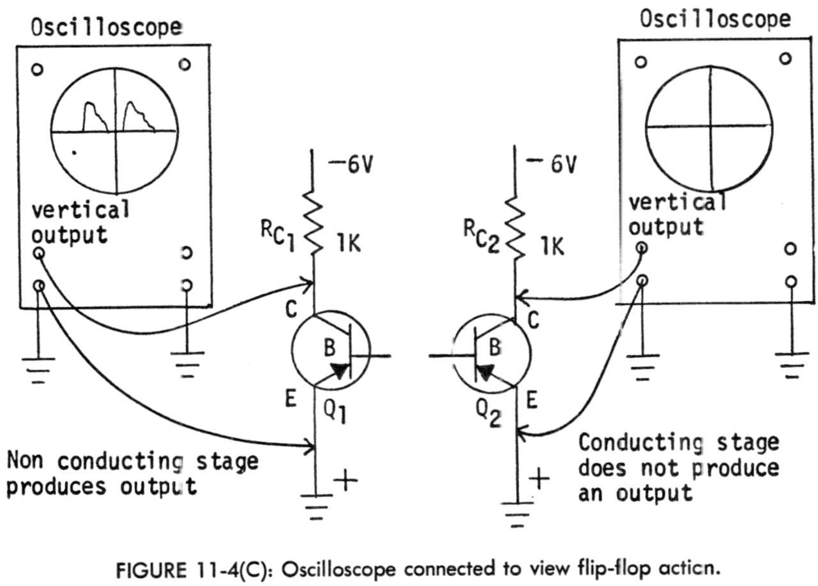

FIGURE 11-4(C): Oscilloscope connected to view flip-flop action.

Now that you have an understanding of how the astable multivibrator operates and what controls the frequency, the following is a circuit and description that you may construct. This circuit shows the possibility of controlling the frequency manually, which is controlling the pulse rate.

Remember, in the formula T = RC, if the value of either R or C is changed, T will also change. Use a variable control to vary the value of R in one of the networks, which will result in a change of time or frequency. Make one of the capacitors large in value and the other one small in value. Since the time will be different in both networks, this will cause the circuit to pulse at a rate determined by the value in the networks. Connect your oscilloscope as shown in the diagram, and view the pulse as the capacitor charges and discharges.

Figure 11-5 is an astable multivibrator in which the frequency can be varied and the time between the pulses can be varied. This circuit may be called a time base generator because the time between the pulses can be varied. The variable resistor, R_{b1}, is the base resistor for Q_2; since it is variable, the forward bias is changed. Varying the resistor causes the capacitor to charge faster or slower, depending on

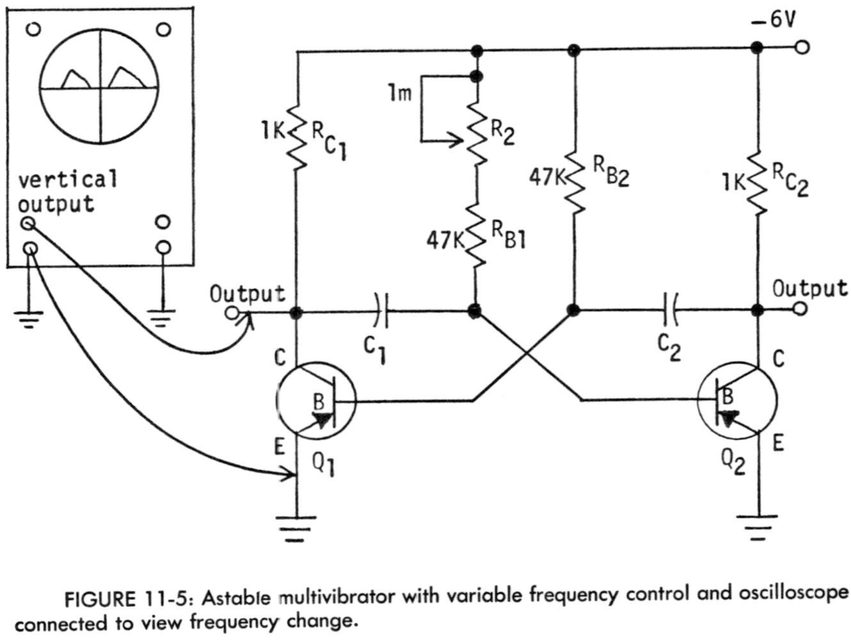

FIGURE 11-5: Astable multivibrator with variable frequency control and oscilloscope connected to view frequency change.

which way the resistance is going at the time. The wave form displayed on the oscilloscope will be a pulse in which the time between pulses can be changed by varying the resistance. The circuit, as well as that in Figure 11-4 and 11-5, could be used as a clock generator to drive a binary counter or as a clock generator for any type of digital equipment.

Figure 11-6(A) shows the astable multivibrator used to flash two small bulbs on and off as the stages switch from one state to another. You can see the stage that is nonconducting when the bulb comes on, and the bulb that is off shows the stage is conducting.

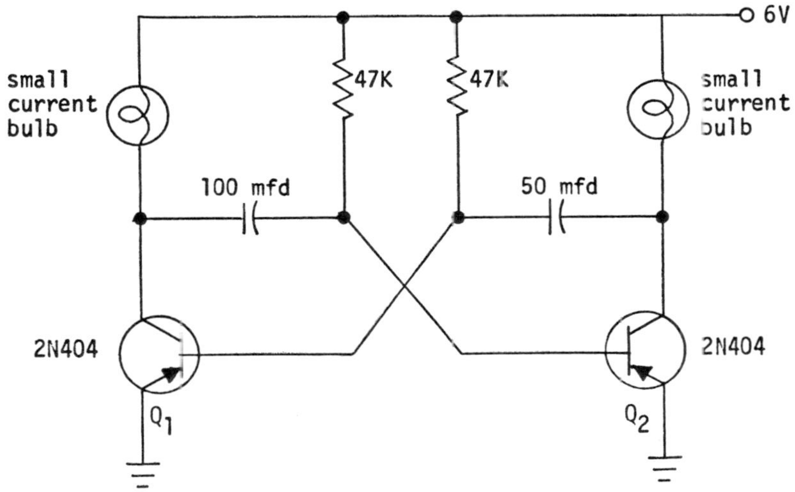

FIGURE 11-6(A): Astable multivibrator with small current bulbs as collector load.

Figure 11-6(B) shows the same circuit, only now the multivibrator is driving an electronic switch to turn off and on a larger bulb with higher voltage and current. Transistors Q_3 and Q_4 are power transistors and should be chosen to handle the voltage and current for the type bulb used.

Both circuits, (A) and (B), could be put to many uses, such as an emergency flasher. They could be mounted on the back and front of a bicycle, one red and the other green. There are many more possible uses; I am certain you can think of quite a few.

To see the theory come to life in a monostable multivibrator circuit, construct the practical circuit shown in Figure 11-7(A). You

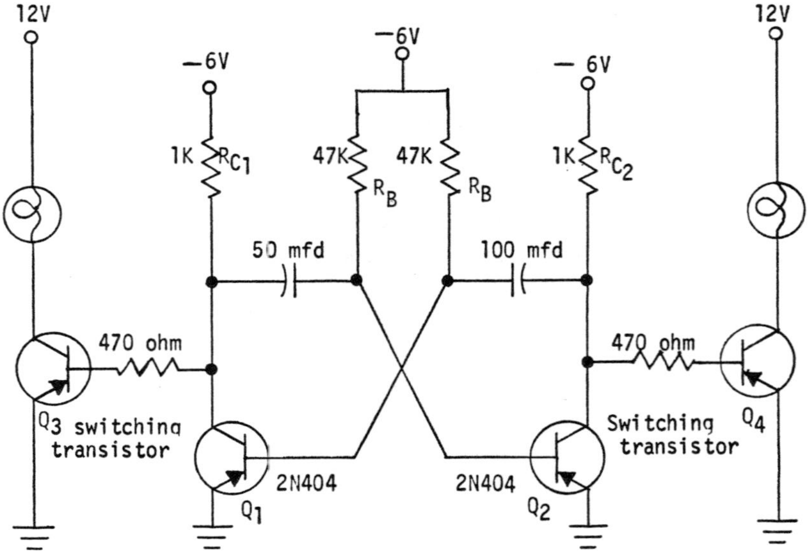

FIGURE 11-6(B): Astable multivibrator driving switching transistors which turn bulbs off and on.

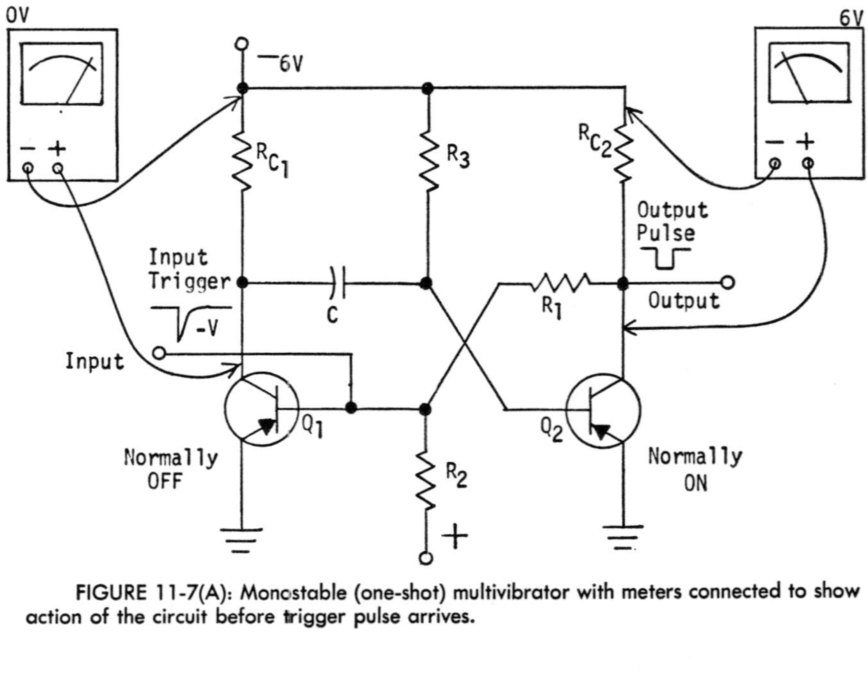

FIGURE 11-7(A): Monostable (one-shot) multivibrator with meters connected to show action of the circuit before trigger pulse arrives.

will use a voltmeter to measure the voltage drop across the collector load resistors R_{C1} and R_{C2}. This will tell you which stage is conducting and which stage is cut off.

When connecting the meter to this circuit make certain of the meter polarity; observe that this is proving the direction of current flow through the PNP transistor. Q_1 is normally cut off. This is because of its base being connected through R_2 to the high positive voltage. This reverse biases the base of this transistor. Q_2 is normally on or conducting; its base is forward biased through R_3. The output of Q_1 is coupled to an outside source which is the trigger pulse that is used to forward bias that stage when it is applied. When this pulse is applied and Q_1 base is forward biased, it will start conducting. C_1 will charge up with a positive voltage when the collector starts drawing current. This positive voltage will cut off Q_2.

In Figure 11-7(A) the condition of the circuit is shown when a supply voltage is applied but before the trigger pulse is applied. The meter across load resistor R_{C1} reads zero, which indicates that it is not drawing current. The meter across the load resistor R_{C2} is reading the full supply voltage, which tells you it is drawing current.

Figure 11-7(B) shows the meters connected to read the collector voltage of Q_1 and Q_2 after the input pulse has been applied. The meter connected to the collector of Q_1 is reading zero, which indicates that this collector is drawing current down through the load resistor and all

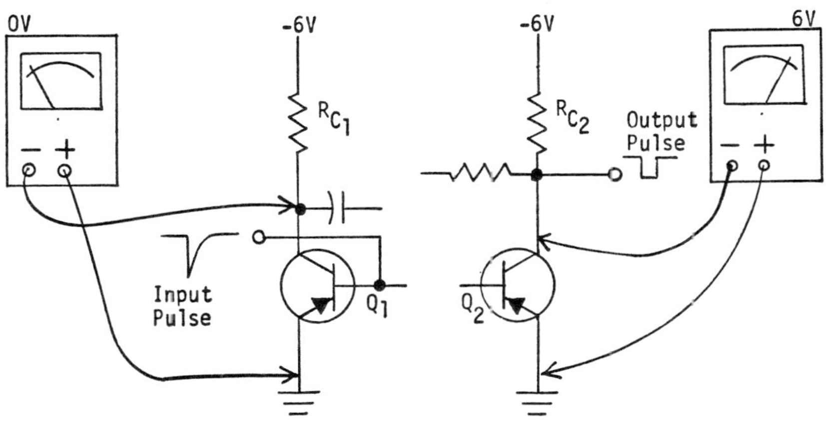

FIGURE 11-7(B): Meters connected to monostable multivibrators to show action after trigger pulse arrives.

of the supply voltage is being dropped across the resistor, with no voltage appearing at the collector. This stage is in the *on* (conducting) condition. The meter connected to the collector of Q_2 is reading the supply voltage. This condition indicates that no current is flowing in the collector from the load resistor and the collector is producing an output. After a short time, which is determined by the CR_3 network, the circuit will return to that shown in (A). The meter has shown the flip-flop condition of the circuit caused by the input pulse.

Figure 11-7(C) shows a method of connecting an oscilloscope to this circuit to view the output pulse after the input pulse has been applied. Notice that the pulse disappears after a short time, indicating the circuit has returned to its original state, that of (A). Another oscilloscope could be connected to the collector of Q_1 to monitor the input pulse. Watch it disappear; then a short time later watch the pulse at Q_2 disappear. This will give you a complete picture of the cycle of flip-flop in the monostable multivibrator.

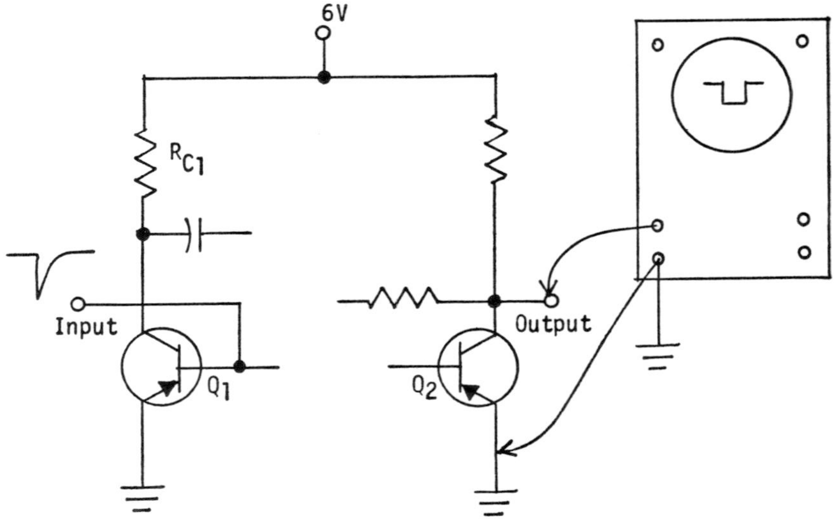

FIGURE 11-7(C): How to connect oscilloscope to monostable multivibrator to view the output wave form after trigger pulse has been applied.

Figure 11-8(A) shows a bistable multivibrator circuit in which meters are connected across the collector load resistors in each stage to

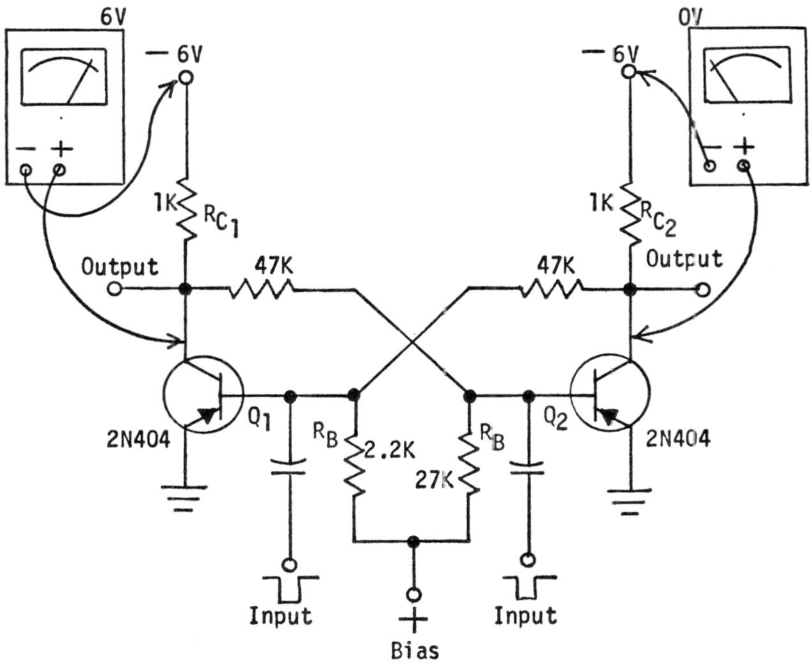

FIGURE 11-8(A): Bistable multivibrator with meters connected to show action of flip-flop after trigger pulse has been applied.

read the voltage drop across them. This reading on the meters will tell you which stage is conducting and which stage is not conducting. Notice that M_1 across R_{C1} is reading six volts; this tells you that this stage is conducting, that current is passing down through the resistor and the full supply voltage is being dropped across it. M_2 across R_{C2} is reading zero; therefore there is no current flowing through this resistor, and this stage is cut off.

Figure 11-8(B) shows two meters connected to the collectors of this circuit to measure the voltage after the input pulse has been applied. M_1 is reading zero; this indicates that the negative pulse applied to the base of Q_1 has forward biased this stage and now it is conducting. The forward bias has caused current to flow down through R_{C1} into the collector. This current flow through R_{C1} has caused all of the supply voltage to be dropped across the resistor, and the full supply voltage is appearing at the collector. This condition could remain inde-

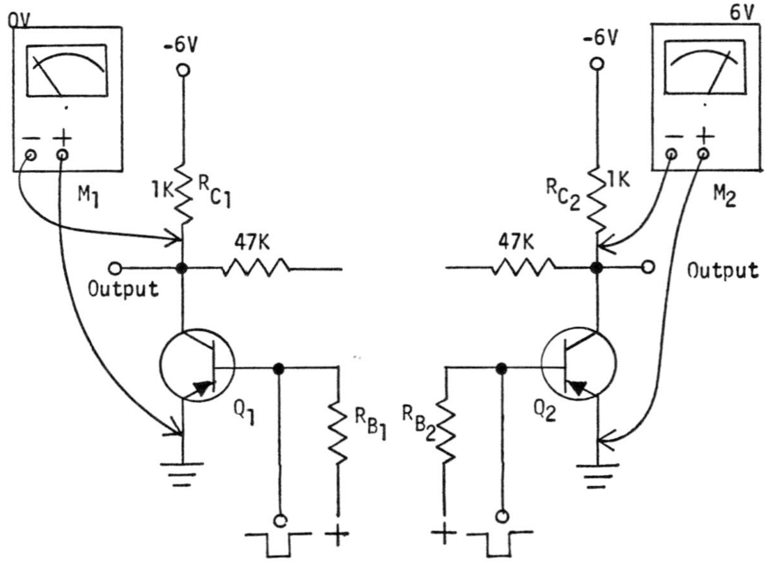

FIGURE 11-8(B): Meters connected to show condition of flip-flop circuit after first pulse has been applied.

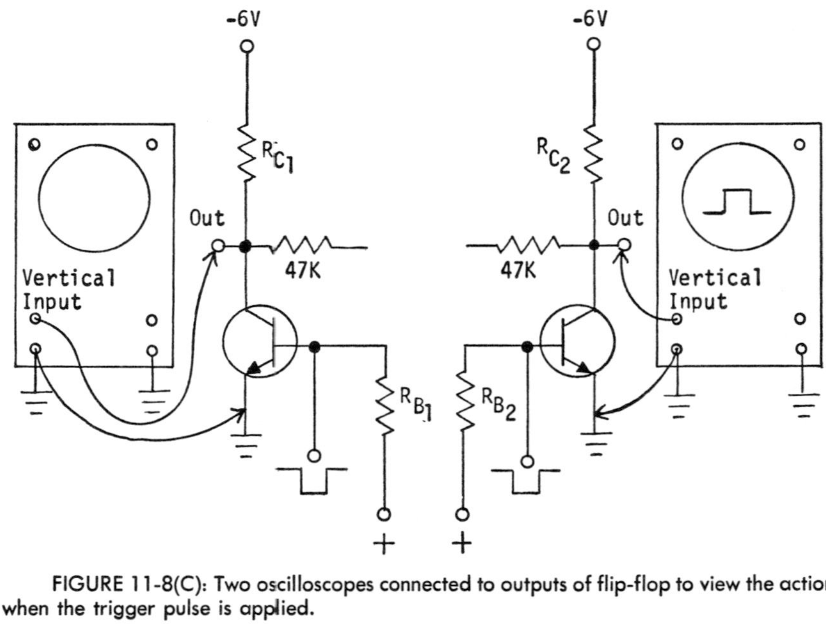

FIGURE 11-8(C): Two oscilloscopes connected to outputs of flip-flop to view the action when the trigger pulse is applied.

finitely or until another pulse is applied to change the state. When the next pulse is applied, the meters at (B) will reverse themselves, and the circuit will return again to that condition shown in (A), completing one cycle.

Figure 11-8(C) shows how to connect an oscilloscope to this circuit to view the circuit operation during the flip-flop states. The condition shown is after the first pulse has been applied. When the second pulse has been applied, the stage will reverse, and the pattern will be shown on the oscilloscope on the left. At this time the right side will not show a wave form, indicating that this stage is conducting and not producing an output. If both oscilloscopes were left attached while the input pulses were being applied, you would see the flip-flop from one oscilloscope to the other.

WHAT THIS CHAPTER HAS TAUGHT YOU

A multivibrator is two inverter amplifier stages coupled in such a manner that the output of one stage is in phase with and drives the other input. This action is positive or regenerative feedback. The circuit's operation depends upon the type of coupling, and the frequency depends upon the values in the feedback network or the trigger pulse fed into the circuit from an outside source.

There are three types of multivibrator circuits: astable, monostable, and bistable. The astable is the only one that is an oscillator circuit, and it is called a "free-running" oscillator. The monostable is called "one-shot," and the bistable is called the "flip-flop" circuit.

You have constructed practical circuits and used proper test equipment necessary for testing these circuits. You have a basic knowledge of circuits used in the most modern and complex electronic equipment today.

12

Television Transmission

VIDEO CAMERA OPERATION

The video portion of a television broadcast actually starts at the camera; here the image to be televised is changed from a light energy to an electrical energy. It is this energy that is processed from transmission at the studio, goes to the transmitter, is sent out as an RF signal, and finally reaches your television receiver. Here it is processed and finally appears on the face of the picture tube in the same form as it did originally.

The word *video* is defined as pertaining to picture signals or to the sections of a television system that carry these signals in either unmodulated or modulated form. Thus, when referring to the video signal in television, we are referring to the picture portion of the signal.

The video camera has a lens, and when the distance and angle from the subject change, the proper lens is selected. The lens focuses the light energy on the face of the vidicon tube. This is the most popular of all the camera tubes in the modern broadcast system, as well as in closed circuit television. The face of this tube is flat, and the inside surface is covered with three very thin films of light active material. The first is a transparent conductive material, the second is a semiconductor photoresistive material, the last is a photoconductive mosaic material. The combination of these three layers of material has the ability to transform the light energy focused on it by the lens, to an electrical energy. This electrical energy, in the form of an electron beam, is focused on the cathode of the tube. It is removed from the cathode and fed to a pre-amplifier stage; from here it is fed to regular video amplifier stages in the camera. In order that this video can be displayed on the face of your CRT (cathode-ray tube) or picture tube in the proper manner, accurately generated sync pulses are added to it before it is transmitted. These sync pulses are the horizontal sync pulse at 15,750 Hz, the vertical sync pulse at 60 Hz, and the vertical blanking pulses, also at 60 Hz.

These sync pulses are combined with the video signal and sent to the visual transmitter.

THE VISUAL AND AURAL TRANSMITTER OPERATIONS

The visual transmitter is an amplitude modulated system. The carrier signal is generated at a low frequency by a crystal oscillator. The crystal oscillator is followed by a series of multiplier stages. Three multiplier stages are used from the oscillator to the final amplifier for the low band (channel 2 to 6), and nine multiplier stages are used for the high band (channel 7 to 13).

The multiplier stages are followed by one or more intermediate power amplifiers; they are followed by a class C amplifier, which is the final stage. This RF carrier is modulated by the video signal.

Figure 12-1 shows a block diagram of a typical visual transmitter and studio stages. Shown also is the sync generator, which is a part of the studio equipment. Shown below the camera and its stages are the wave forms of the video information before it is mixed with sync pulses. The sync generator shows the basic stages with their wave forms directly above them. These wave forms are not drawn to scale, nor are they divided into time intervals. We inserted these wave forms only to show you how complex the video signal is and how much information is contained in the video transmission. The transmission of the picture employs a process known as *vestigial sideband transmission.*

The term *vestigial sideband* is defined as the transmitted portion of an amplitude-modulated sideband that has been largely suppressed by a filter having a gradual cut-off in the neighborhood of the carrier frequency. The other sideband is transmitted without much supression. The word *vestigial* comes from the word *vestige*, which means a trace mark or sign of something which has passed away. Transmission by vestigial sideband means that the complete video information is transmitted on the upper side of the carrier, while on the lower side of the carrier, only the lower frequencies are transmitted, and the higher

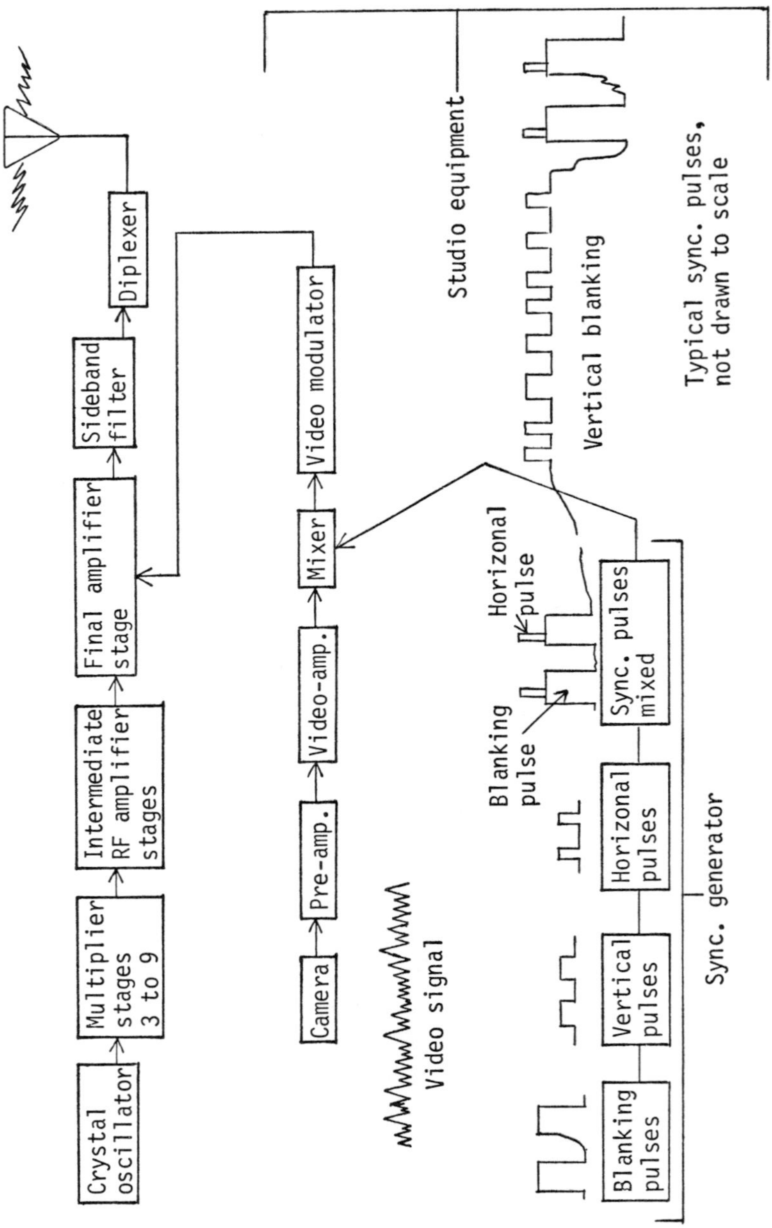

FIGURE 12-1: Typical visual transmitter, and studio stages.

frequency components are surpressed by the use of a sideband filter. Since only part of the frequencies below the carrier are transmitted, the lower sideband extends only to 0.75 MHz below the carrier. The upper sideband extends 4 MHz above the carrier. The system allows a video signal as high as 4.75 MHz to be transmitted, although the assigned band width is only 6 MHz wide.

The picture information transmitted from a visual transmitter is referred to as a *composite video signal*, defined as the signal obtained by combining a monochrome television picture signal with the horizontal and vertical blanking and synchronizing signals. In the following chapter, which deals with television receivers, I shall be using this term. It will mean that I am referring to the complete video and sync pulses combined.

Figure 12-2 shows a composite video signal with horizontal and vertical sync pulses added. This is not an exact scale of this wave form but an example of the combined wave form of the complete visual signal, parts of which were shown you in Figure 12-1.

The aural transmitter in the complete television system is a separate transmitter, using the mode of FM for transmitting the sound portion of the signal. This FM transmitter was described in Chapter 8. If you have forgotten any of the theory of FM transmission, go back to Chapter 8 and review how the FM signal is transmitted.

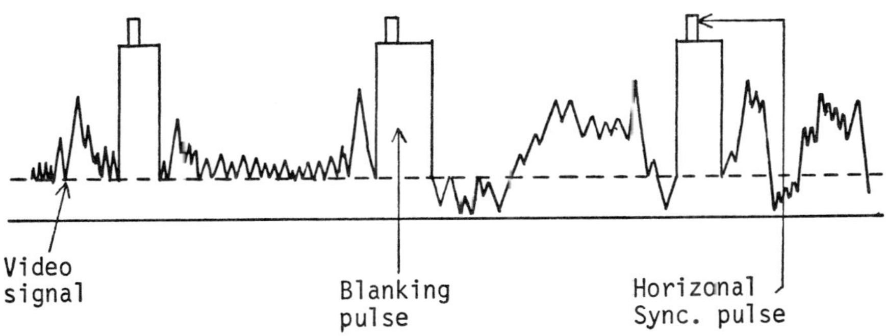

FIGURE 12-2: Typical composite video signal.

TRANSMITTING THE TELEVISON SIGNAL

I have explained that in the television transmitting system there are two transmitters. One is the visual transmitter and the mode of transmission is amplitude-modulation, by a process called vestigial sideband. The aural transmitter is an FM transmitter. Both of these transmitters are coupled to the same antenna through a diplexer. A diplexer is a coupling system that allows two different transmitters to operate simultaneously or separately from the same antenna.

The diplexer provides the proper impedance match between the aural and visual transmitters and the antenna. It is usually designed so that there is no interaction or cross modulation between the picture and the sound. Figure 12-3 shows a diagram of a typical diplexer that might be used at the TV station. The feed-line to the antenna is connected to the output of this unit.

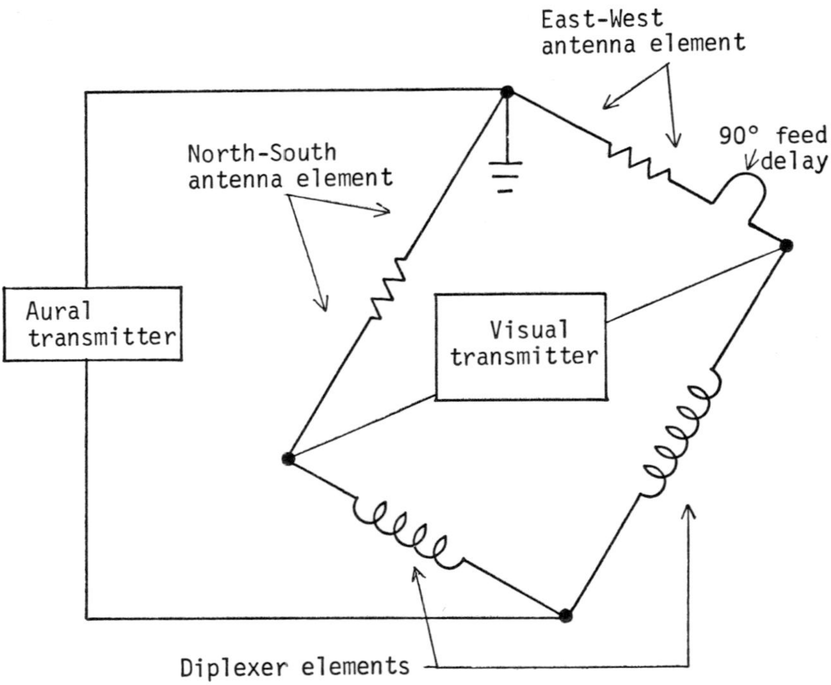

FIGURE 12-3: Typical diplexer unit showing two transmitters connected to one antenna.

Since television transmission is in the VHF (very high frequencies) and the UHF (ultra high frequencies), tne transmission from the antenna is thought to be only in the line of sight. Therefore, it is customary to locate the antenna as high above the surrounding terrain as possible. The antenna is supported on top of a tower that extends to a height of a few hundred feet to around 2063 feet. This will put the antenna as high as possible so that the station will cover as much viewing audience as possible. Television stations run much more power to their final amplifier stage than almost any other type of service. The antenna is so constructed that it also has considerable gain in radiating this power.

Probably the greatest advancement in television viewing was the introduction of CATV and relay systems. CATV or *community antenna television* brought to many cities and communities the pleasure of television viewing even though they were considered to be in the fringe area. The fringe area meant they were not in the line of sight of the transmitting antenna or were located in a valley where the surrounding terrain cut off the signal from the transmitting antenna. The CATV system located their receiving antenna on the line of sight from the transmitting antenna. The signal was amplified and transported by co-ax cable, in some locations for miles, into the city or town. Then it was distributed by co-ax cable to the homes of the viewers. In some cases these systems may use up to two or three hundred miles of cable and fifty to one hundred amplifiers, but they do help to extend the range and coverage of many television stations.

Today these systems are in our largest cities in which there may be four or five television stations; with CATV viewers also have access to stations from other cities that they would not see otherwise. Figure 12-4 shows a typical CATV system that may be used where the terrain is very irregular and where it would be impossible to receive a TV signal without the aid of this service.

Before long, CATV will add new services; one of these may be a tie-in with a ground receiving station from one or more of the communications satellites. This will give the viewer an opportunity to tune in on TV programs from any of the foreign countries served by the satellite. Another possible service in the near future will be that the system will be made two-way. In this arrangement the system will be tied into a central computer, and you will have a telephone similar to the one now being used in your home. You will pick up the phone, dial a number, and ask the computer for certain information. In a matter of minutes the answer will be written across the face of your TV screen.

There are many other services that could be rendered by the CATV system, and we may see some of them in the very near future.

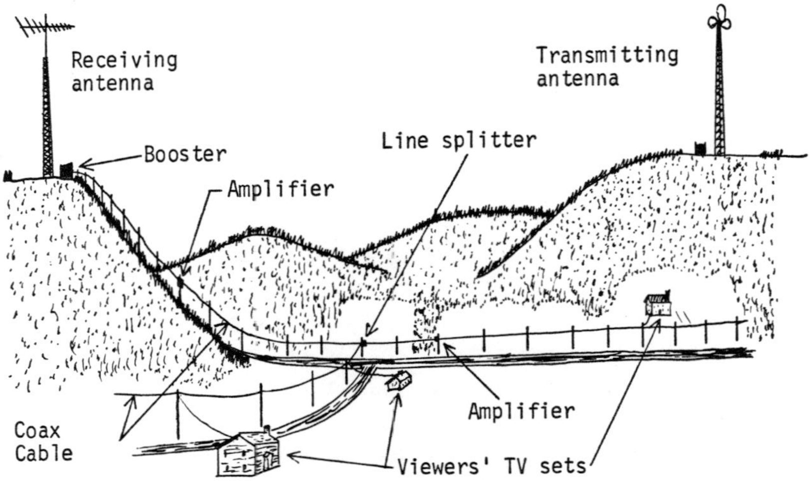

FIGURE 12-4: Typical CATV installation.

13

Television Receiver Circuit Analysis

"These receivers are not like they were a few years ago." This is what the technicians who have been in the business for a few years are saying. We have to agree; they certainly are not like the older sets of a few years back.

Most modern television receivers are all or part solid-state. Some use the hybrid circuits; that is, they have transistors and some vacuum tubes. Others use transistors and IC chips. Very few have a chassis, printed circuit boards, or modules that plug into a frame. Many of these circuit boards contain two or more circuits and are encapsulated; that is, when they are made, they are dipped into a thick protective coating of plastic fluid, which dries very hard and cannot be removed. Therefore, the module cannot be repaired when removed but is discarded.

Many of the modern receivers have very elaborate switching circuits, both manual and automatic. The remote control circuits are designed for changing stations, adjusting volume, turning the receiver off and on, and so forth. These circuits look like mini computers. One receiver came to the market just recently with a circuit that has a digital read-out that gives you the time and the day of the week in the corner of the TV screen.

These are but a few of the extras on your television receiver today, and there are quite a few more to appear in the very near future. How about a television receiver that has a built-in video tape recorder? If you must be away from home and miss your favorite program, before you leave you program the receiver to come on, record your program, cut out the commericals, and turn itself off. When you come home, you can play back your favorite program at any time you wish.

The illustrations on the following pages are some of the latest circuits that may be used in the modern television receiver.

In Figure 13-1, I have presented a block diagram of the circuits as they would appear, showing their function in a black and white television receiver. The arrows indicate the path of the signal, how the composite signal is divided at certain points, and how that portion is directed to the location where it is to be used.

Before we enter into a discussion concerning the circuits, let us review some facts concerning the signal.

In commercial television there are twelve channels in the VHF (very high frequency) range and 70 channels in the UHF (ultra high frequency) range.

The VHF range is divided into two different segments, the low band and the high band. The low band begins at 54 MHz (the lower end of channel 2) and extends to 88 MHz (the high end of channel 6). The high band begins at 174 MHz (lower end of channel 7) and extends to 216 MHz (high end of channel 13).

The UHF band begins at 470 MHz (lower end of channel 14) and extends to 890 MHz (high end of channel 83). Each channel is 6 MHz wide in both the VHF and UHF bands.

A television receiver is almost like two receivers in one. The RF signal at the antenna terminals contains an AM (video) signal and FM (sound) signal information. Both of these signals, along with other information, are contained in one signal that is referred to as a *composite television signal*.

THE RF TUNER

The tuner, as it is referred to, is the input to the television receiver. It contains an RF amplifier, a local oscillator, and mixer stage. There are two tuners to each receiver, one for the VHF and one for the UHF. The only difference between them is the frequency at which they operate.

Turn now to Figure 13-2. This is a typical broadband RF amplifier (we say broadband because it must amplify a band of frequencies from 54 to 216 MHz), or the complete VHF band. The input transformer, or balun, matches the 300 ohm input to the low impedance at the base of the transistor. Notice that a NPN transistor is used. Remember, this transistor will operate better at the higher frequencies because of the

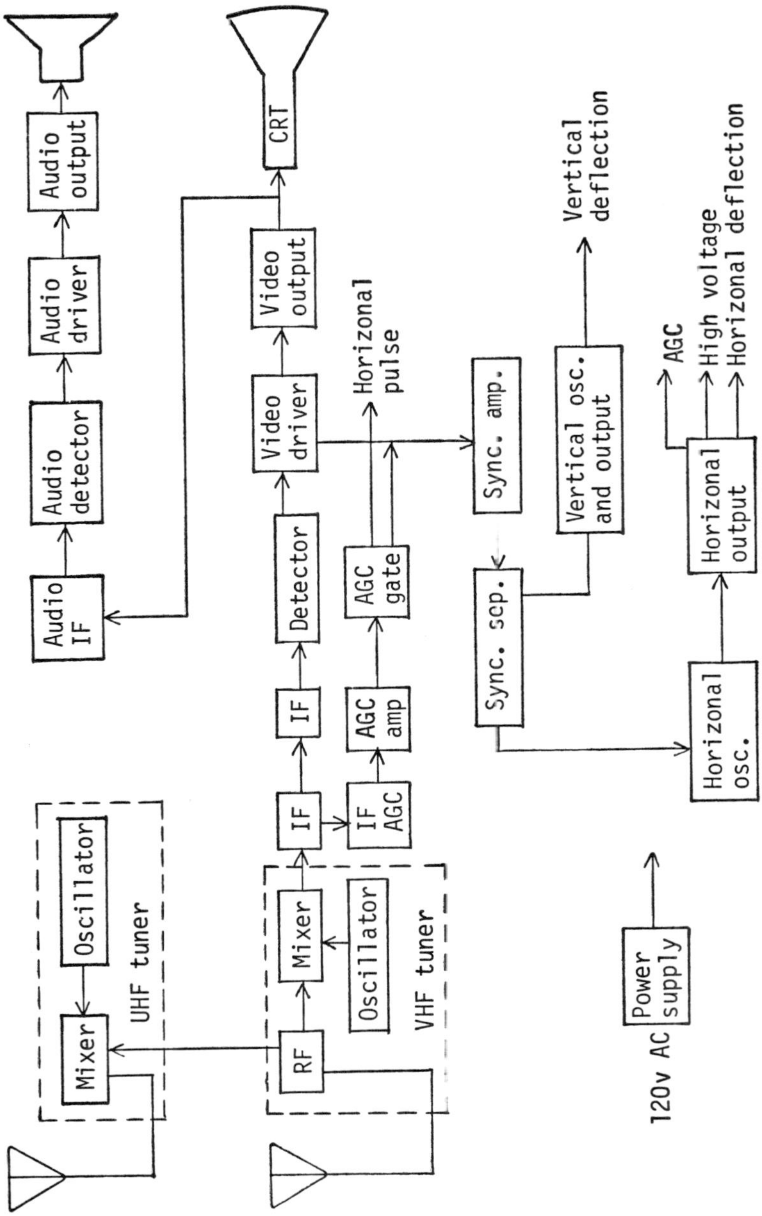

FIGURE 13-1: Block diagram of television receiver.

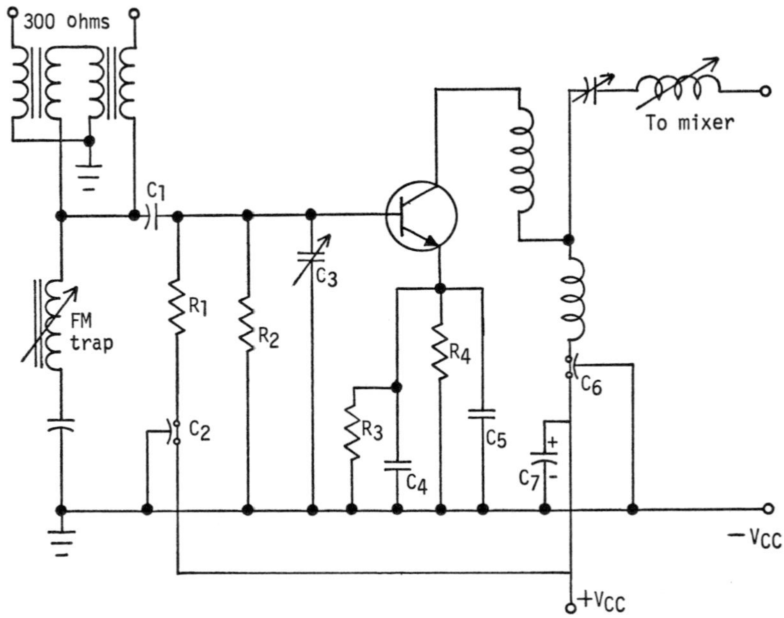

FIGURE 13-2: Broadband RF amplifier that may be used in tuner of a television receiver (50-450 MHz).

transit time of electrons. The circuit is a common emitter. The output is connected to a broadband transformer T, and the output of this stage is coupled to the mixer stage.

Turn now to Figure 13-3. This is a typical mixer and oscillator stage in the tuner. The heterodyning action takes place in this stage by the same process as it did in the AM and FM receivers.

One of the unusual or different things you might notice about this circuit is the use of the cascade mixer stage. It is claimed that this circuit will eliminate the collector to base feedback capacitance. A common-emitter first stage operates into a low input impedance second stage. This makes it unnecessary to neutralize the mixer because of the isolation between the base of the mixer and the output of the second stage.

The oscillator circuit uses a split inductor-capacitor combination feedback between the collector and the base. The oscillator coil is mounted on the channel selector switch, and a different coil is switched into the circuit when the channel selector is rotated.

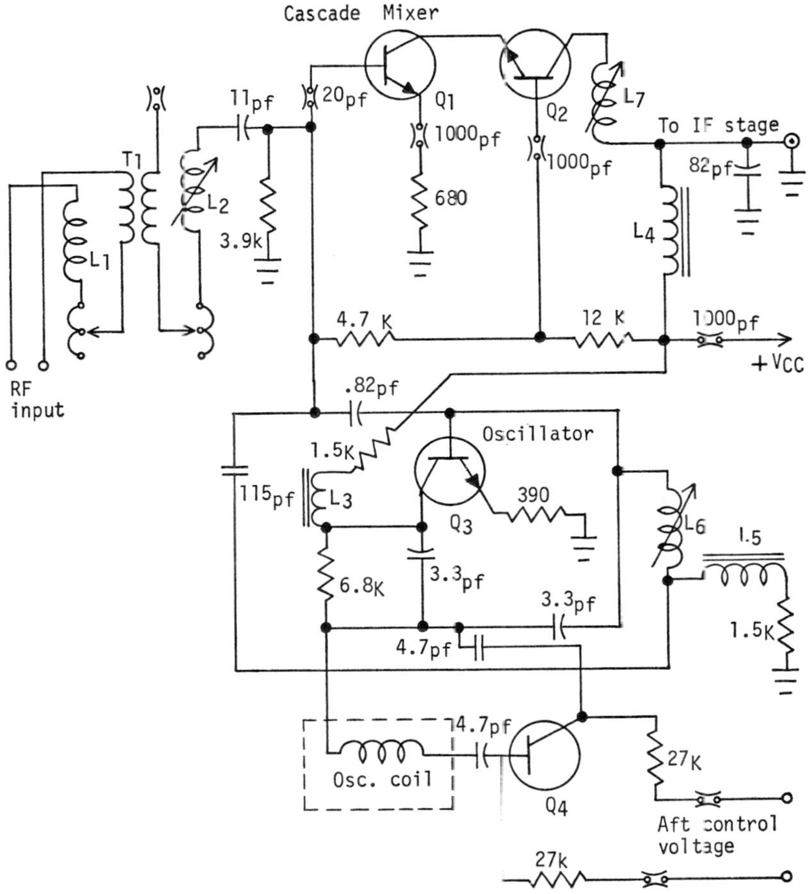

FIGURE 13-3: Typical mixer and oscillator circuit.

One other part of this circuit that you will find different from the older receivers is the AFT (automatic fine tuning) circuit. This is Q_4 with only the base-collector junction connected as a corrective diode.

The capacitance formed by this junction is placed across the oscillator coil. It is made variable by the application of the AFT voltage, and the increasing or decreasing of this capacitance will lower or raise the oscillator frequency automatically. The oscillator voltage is fed into the cascade mixer stage and heterodyned with the incoming RF signal, and the IF signal is removed from the collector.

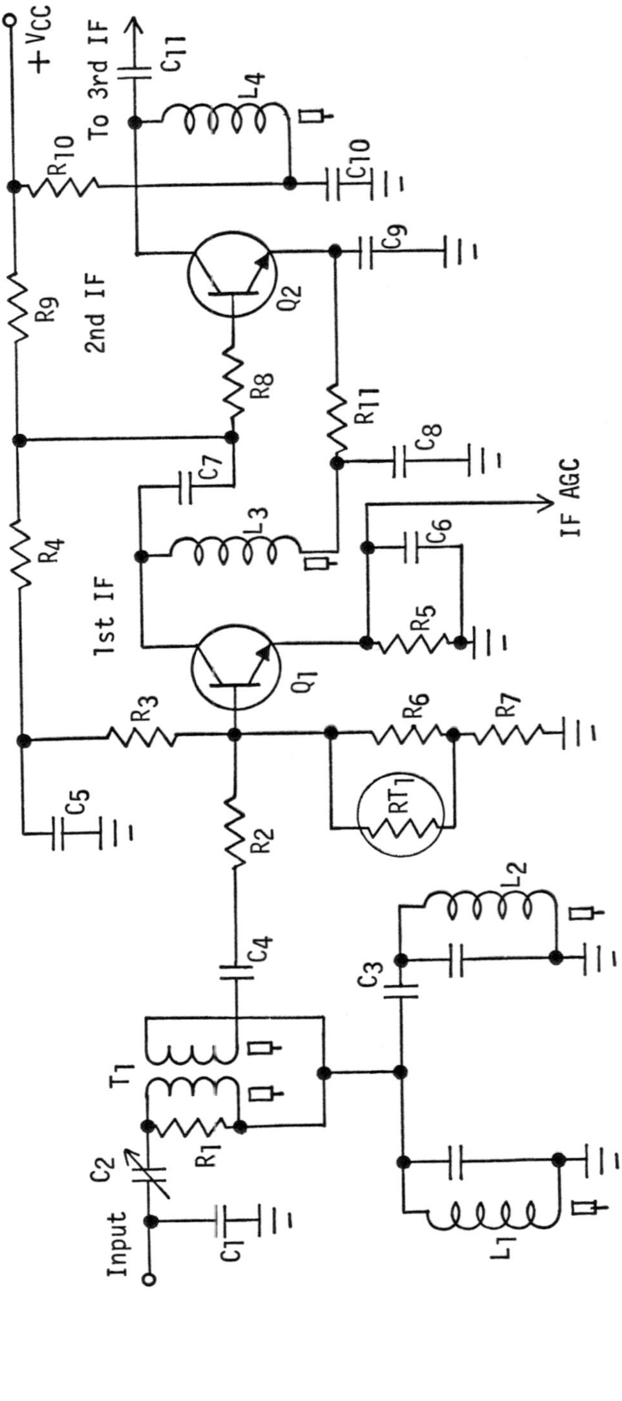

FIGURE 13-4: Typical first and second IF stages.

INTERMEDIATE FREQUENCY

Refer to Figure 13-4. The IF signal is fed to the first IF stage through T_1 and coupled to the base of this stage. L_1 is a 47.25 MHz sound trap, while L_2 is a 41.25 MHz sound trap. The first stage, Q_1, is impedance coupled to the second stage, Q_2 by L_3. These two stages are staggered tuned; that is, L_3 is tuned to 45.75 MHz, and L_4 in the collector circuit of Q_2 is turned to 42.17 MHz.

AGC (automatic gain control) voltage is applied to the first stage, Q_1, at the emitter. This stage then controls the gain of the second stage, Q_2. These two transistors are series connected to the DC supply. We can trace the currrent flow to make sure. Follow me as I take this current path. As resistor, R_5 is common to both transistors. The current starts here at the negative end of this resistor. It moves up through the emitter, across the base and out through the collector; down through L_3 and across R_{11} and up through the emitter of Q_2; across its base into the collector and through L_4, through resistor R_{10} to the positive side of the supply.

Thermistor RT_1 in the base of Q_1 prevents temperature changes from affecting the gain of the IF stage. The signal is removed from the collector of Q_2 and fed to the third IF stage.

THIRD IF AND VIDEO DETECTOR

Refer to Figure 13-5. This is a typical IF and video detector stage. The signal is fed to the base of Q_1; here it is amplified and the collector is connected to a double-tuned transformer. This consists of L_1 and T_1; L_1 is inductively coupled to the secondary of T_1. Coupling between L_1 and T_1 is by C_{10}, and the secondary of T_1 is coupled to a diode video detector. L_2 at the bottom of T_1 is a sound trap tuned to 41.25 MHz. C_9 couples the sound to the sound stage.

AUDIO STAGE

A typical audio module found in the modern television receiver is shown in Figure 13-6. Coil L_1 is the sound take off, and the 4.5 MHz signal is coupled to pin 1 and 2 of the IC. L_2 is the quad coil, and it is coupled to pins 9 and 10 of the IC.

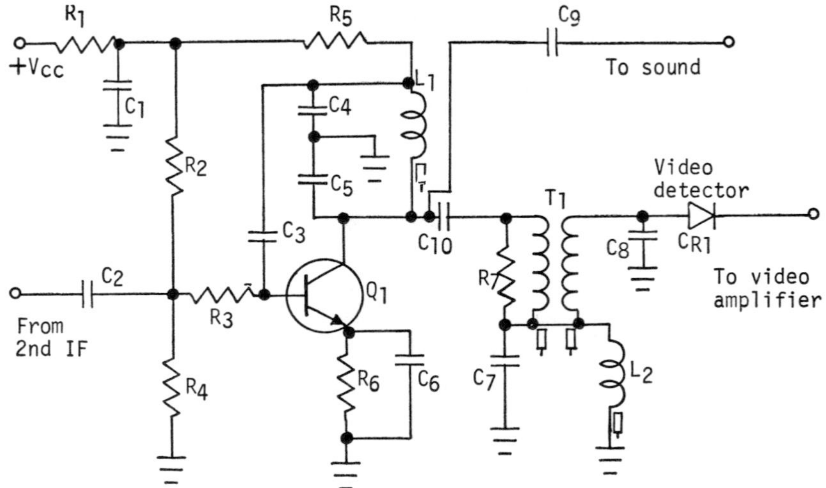

FIGURE 13-5: Third IF section.

The output of the IC, which is terminal 12, is capacitive coupled to the driver stage, Q_1. Notice where the volume control is in this stage; it acts as a variable voltage divider for the voltage developed inside the chip. There is no signal on the control. Q_1 is DC coupled to the output stage which is a complementary-symmetry circuit.

Notice that there is no output transformer; the signal is capacitive coupled to the two speakers and they have an impedance of 16 ohms. One precaution about this type of output: if you have the speakers disconnected, make sure that you reconnect the same coded wires in order to polarize the speakers.

VIDEO STAGES

Figure 13-7 shows a video IF output and video preamplifier stage that may be used in the modern television receiver. T_1, along with L_2, is tuned to 41.25 MHz; the output is then coupled to the input of the IC_1. This chip amplifies the IF, produces the 4.5 MHz sound signal, and amplifies it for the sound stage. The video is also amplified in the IC and coupled through the 4.5 MHz sound trap to the video preamplifier, Q_2. Q_1 acts as a voltage regulator; base voltage for this stage is

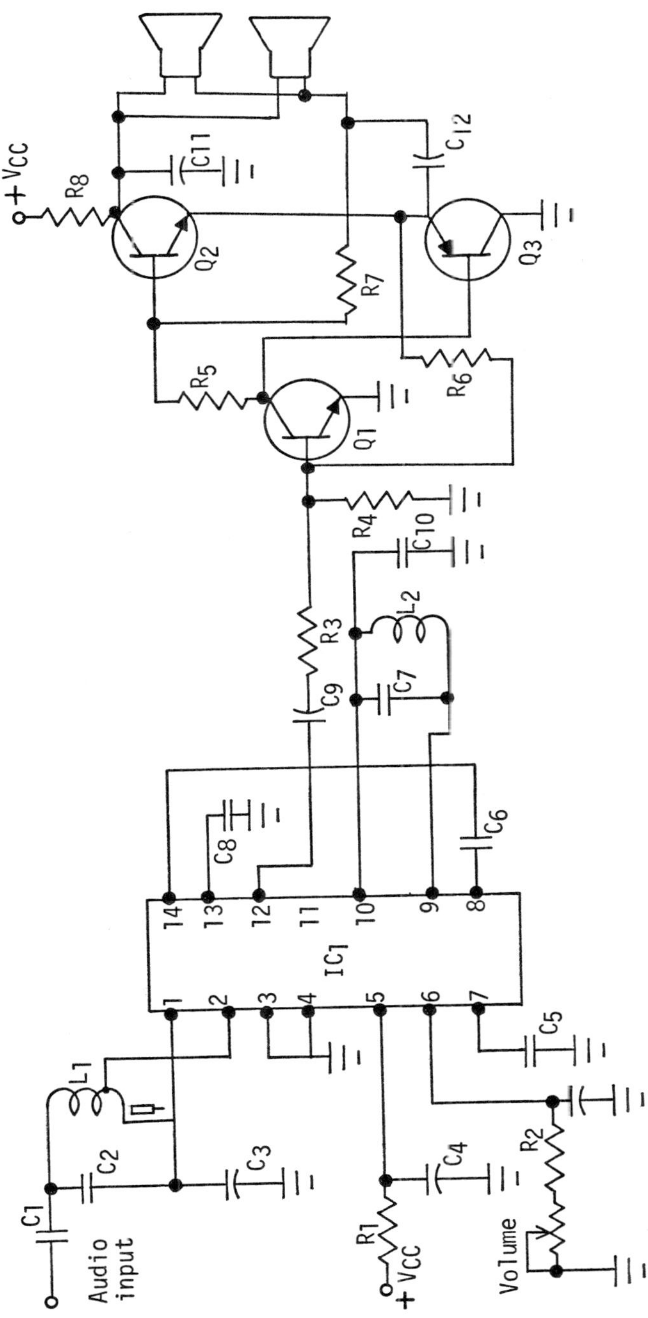

FIGURE 13-6: Typical audio stage.

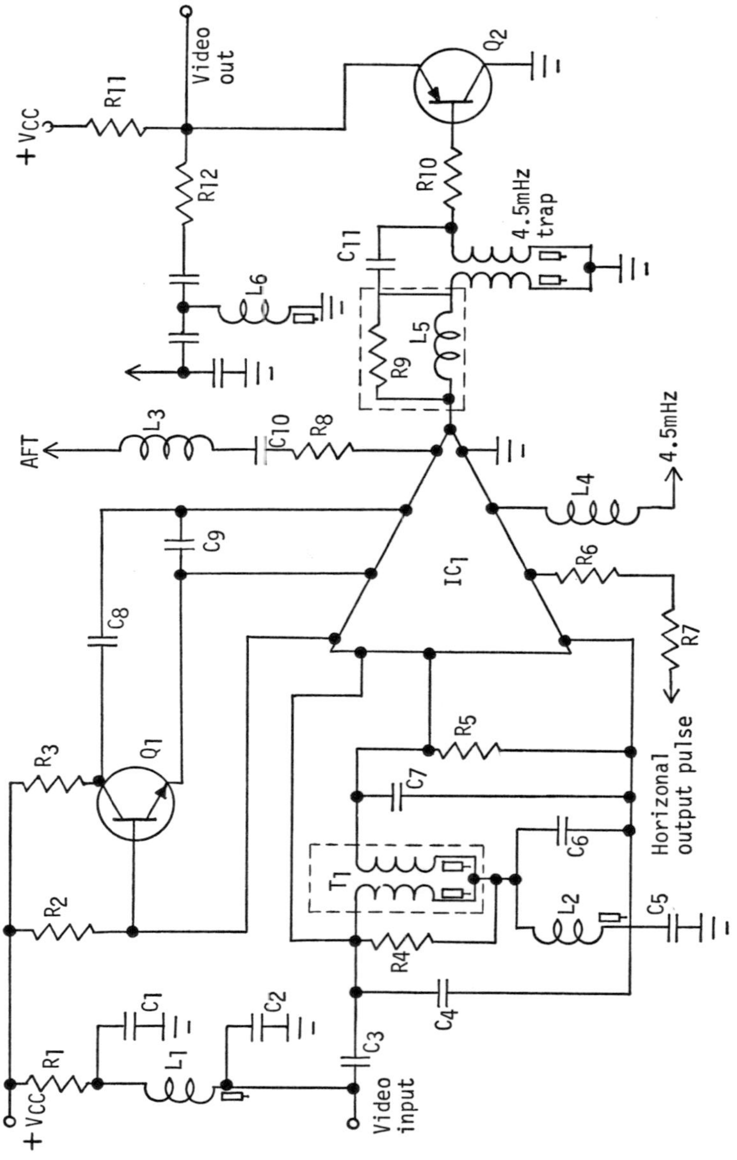

FIGURE 13-7: A possible circuit that could be used as IF output and video preamplifier.

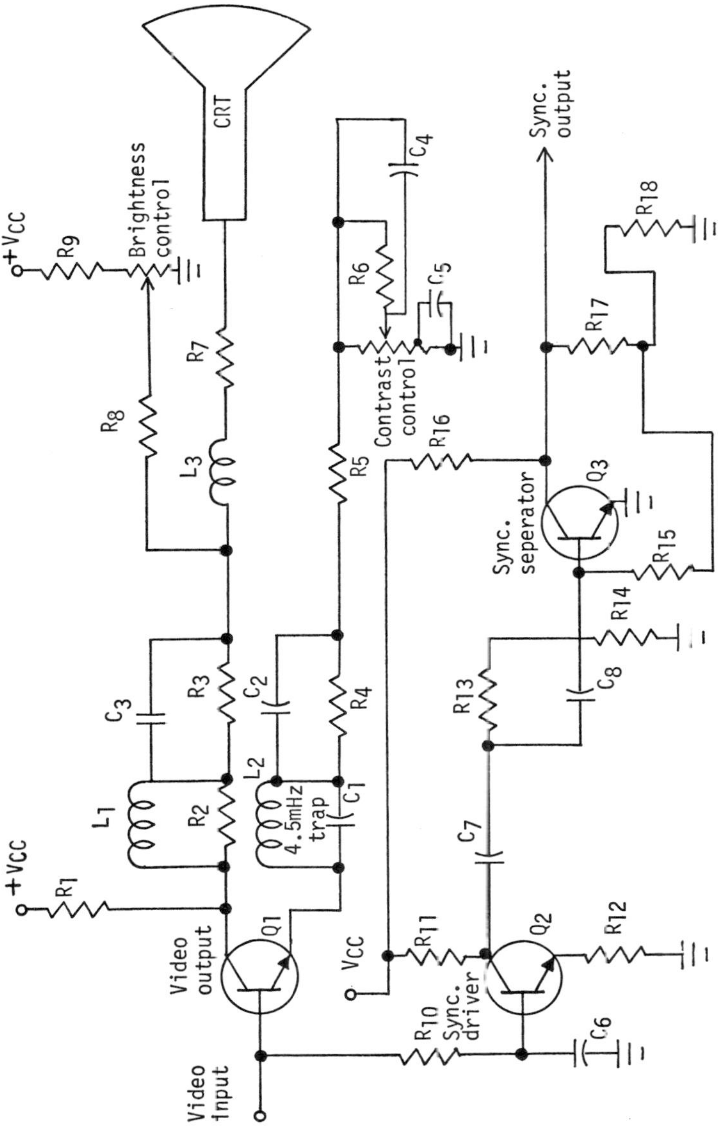

FIGURE 13-8: Video output and sync circuitry module.

established by a zener regulator in the IC. The preamplifier, Q_2, amplifies the signal and couples it to the video output stage.

VIDEO AMPLIFIER OUTPUT, CATHODE RAY TUBE AND SYNC

Figure 13-8 shows a video output amplifier and sync circuit combined as one module. The upper portion of this diagram contains the video output and the lower part is the sync driver and separator.

The video signal is applied to the base of Q_1, the video amplifier. L_2 and C_1 is a parallel tuned circuit at 4.5 MHz in the emitter circuit, controlling the gain of this stage. The collector of this transistor is coupled to the peaking network L_1, R_2, R_3 and C_3, which couples the video to the cathode ray tube cathode.

The input video signal is also coupled to the base of the sync driver Q_2. The signal is amplified in this stage and coupled to the sync separator through C_7. In this stage the signal is separated into horizontal and vertical pulses and applied to their respective stages.

AGC CIRCUIT (AUTOMATIC GAIN CONTROL)

There are many varieties of AGC circuits used in modern television receivers. Since most receivers are all solid-state, many of these AGC circuits use a combination of transistors, diodes, and IC chips with built-in AGC. Figure 13-9 shows one type of circuit that could be used in the AGC. This circuit consists of three different stages. The AGC gate receives the video information into its base while the collector receives negative pulses from a special winding on the horizontal output transformer. These pulses turn on the gate as the incoming sync pulses arrive at the base. The conduction of the gate and the development of the AGC voltage are controlled by the amplitude of the sync pulses contained in the video signal. The diode in the collector prevents the collector-base junction from being forward biased. The small voltage on the emitter prevents the transistor from conducting until the incoming sync pulse drives the base into conduction. With no incoming signal, Q_2, the first AGC amplifier, is cut off because it also is connected to the winding of the horizontal output transformer.

At this time there is no voltage drop across R_4. Therefore, the base of the second amplifier, Q_3, is forward biased and is conducting

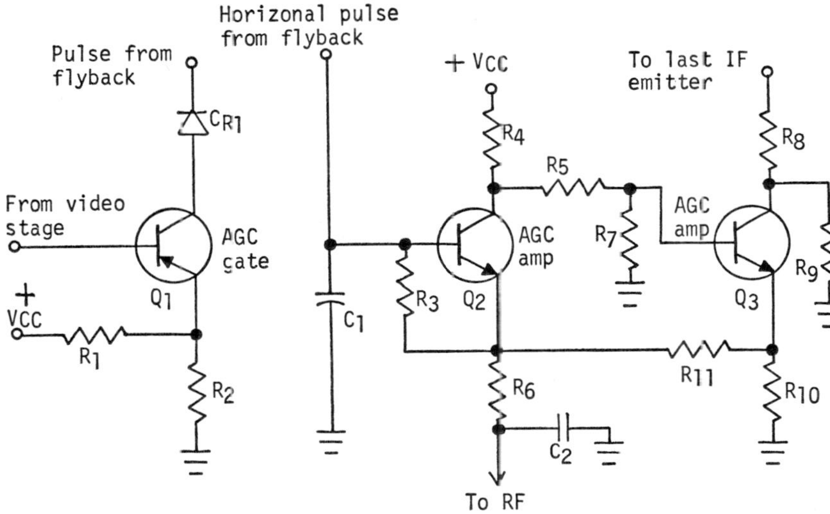

FIGURE 13-9: Possible AGC circuit.

heavily. Since it is connected to the emitter of the IF amplifier, it also draws current and its gain is maximum.

The emitter of the first AGC amplifier is connected to the RF amplifier. The small voltage developed across the emitter resistor is applied to the RF stage, which causes its gain to go to maximum. AGC delay is developed across the resistor feedback network of R_5 and R_7.

VERTICAL CIRCUIT

Each year a new television receiver appears on the market. There are a few new circuits incorporated in one of the sections of the receiver. This is simply a case of doing the same old thing in a different way. The following circuit in Figure 13-10 is one example of this, and shows the complete vertical circuit built into one module.

At the beginning of the vertical trace the sweep capacitor, C_3, develops a charge; it is connected to the yoke. This charge is coupled to the base of the pre-driver Q_2, and starts it conducting. The pre-drivers are connected in cascade, and the conduction of Q_2 causes the others to start conducting, which in turn drives the output into conduc-

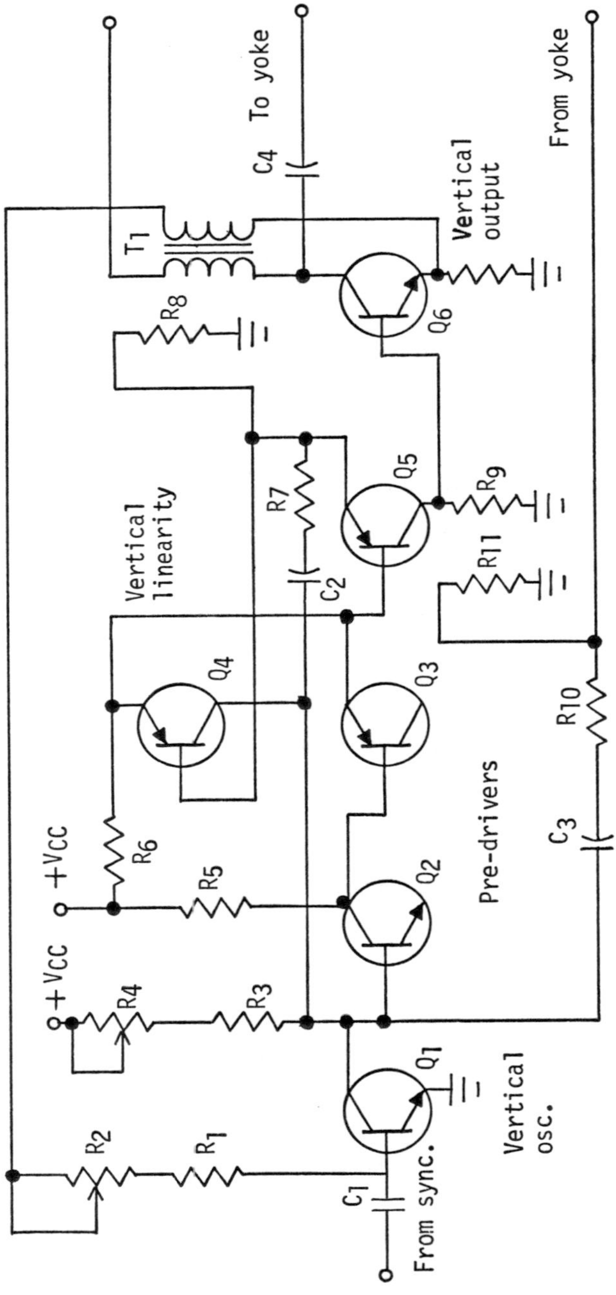

FIGURE 13-10: Vertical oscillator, driver and output.

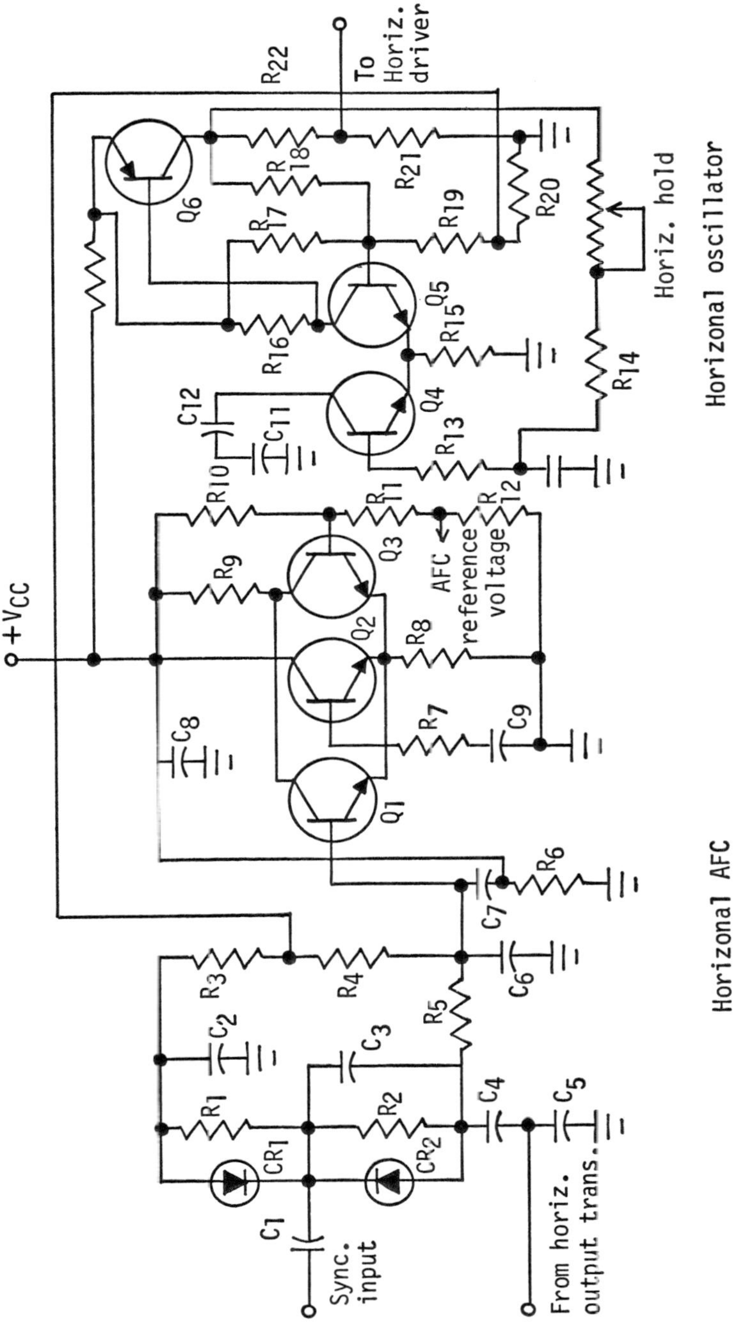

FIGURE 13-11: AFC and horizontal oscillator circuit module that may be used in television receiver.

tion. With the output conducting, a deflection current is produced in the output; this produces the vertical trace.

Feedback is coupled from the secondary of the output transformer T_1, back to the base of the vertical oscillator. This sustains oscillation and R_4, which is the vertical hold control, controls the feedback and thus the frequency. Sync pulses from the sync separator are coupled to the base of the oscillator; these pulses synchronize the oscillator freqency with that of the incoming signal. The trace and retrace of the vertical scanning is caused by the charging and discharging of the sweep capacitor, C_3. During the time the output is cut off, the vertical linearity transistor, Q_4, will conduct; this causes a low resistance path for the C_3 to charge rapidly. This charge on C_3 will cause the predriver to be forward biased and start conducting again. This removes the bias from the vertical linearity transistor and it cuts off. This has produced one cycle of vertical scanning.

AUTOMATIC FREQUENCY CONTROL AND HORIZONTAL OSCILLATOR

In Figure 13-11, I have shown you a modern version of a horizontal oscillator in one section, and in the AFC section, a more modern method of developing AFC voltage. This circuit might be used in the latest television receiver.

Let us look at the horizontal oscillator first. You may not initially recognize this circuit, but it is a form of astable multivibrator. Transistors Q_4 and Q_5 do the switching while Q_6 operates along with Q_5. The frequency of this circuit is determined by capacitor C_{10} at the junction of the voltage divider network of R_{13} and R_{14}. Some control of the frequency is also provided by the horizontal hold control between R_{14} and the collector of Q_6. This frequency is stabilized by the AFC network. Turn to the AFC circuit. The sync pulses are fed to the center of CR_1 and CR_2 comparator network. This part of the circuit is the same as is found in the older model receivers. Notice that a positive DC voltage is applied to this network at the junction of R_3 and R_4 and is removed from the voltage divider in the base of Q_5 in the oscillator section. This voltage raises the reference level above zero. Notice that the collector-emitter junction of the transistor of Q_1 is in parallel with the collector-emitter junction of Q_3. This provides a shunt path. The frequency of oscillation is controlled by increasing or decreasing the time required for capacitor C_{10} to reach the turn-on potential of Q_4. The AFC output and the additional reference voltage are applied to the base of transistor Q_1. When Q_1 starts conducting, this will control the time C_{10} begins to charge and this will control the frequency.

HORIZONTAL OUTPUT AND HIGH VOLTAGE CIRCUITS

Figure 13-12 shows a circuit or module that includes both the horizontal output and the high voltage rectifier circuits. The voltage from the horizontal is transformer coupled to the base of the horizontal output Q_1. The output from this stage is coupled to the horizontal output transformer. Notice the fuse in the collector circuit used as protection device in case of an overload in the output circuit.

The keyed AGC pulse is supplied from one of the windings on the output transformer. The second winding on the secondary winding supplies the retrace suppression. The top end of the primary winding is connected to the high voltage diode which supplies the high voltage for the anode of the picture tube. VR102, CR_1 and CR_2 form a voltage regulator network.

This module could be used in the modern television receiver.

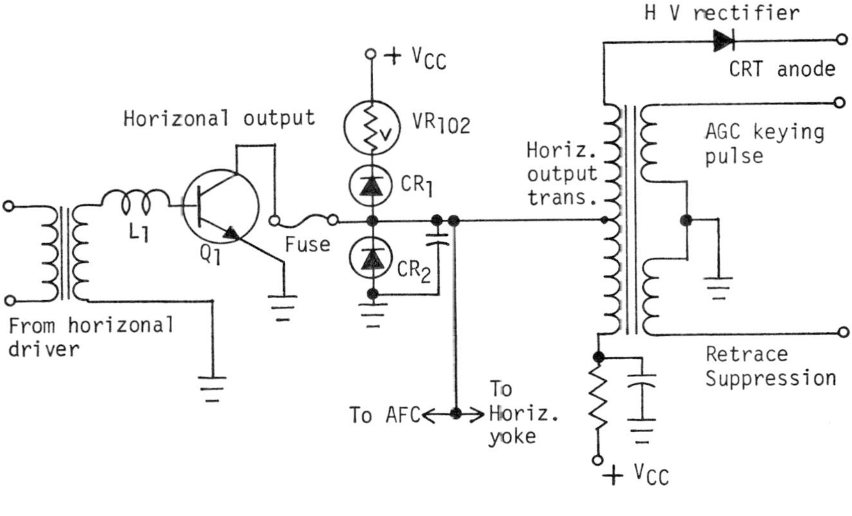

FIGURE 13-12: Horizontal output and high voltage circuit that may be used in television receiver.

The circuits I have shown and briefly discussed are but a few of the new modular circuits used in television receivers today. Since most of these appear in modern electronic equipment, I would advise you to study these circuits and others and apply the theory that you have learned in understanding them.

When you are confronted with trouble in any of these circuits in the future, you will have a better understanding of their operation—a knowledge which will make troubleshooting much easier.

14

Testing Procedure and Shortcuts in Troubleshooting Electronic Equipment

KNOW WHERE TO START

When you are in trouble, the first thing you do is *think*. Then you call on all of your knowledgeable resources and plan the fastest, easiest, and most profitable escape route.

For example, when you are driving down the highway in your car and you have a flat tire, would you get out and raise the hood to start the repair of this trouble? Certainly not. You would look at all the tires to determine which one was flat. By visual inspection, you localize the fault. You decide what equipment is necessary to remedy this fault and the quickest way to use this equipment. Next you decide what part or parts are necessary for the repairs; then you proceed with the repairs.

You have used a systematic approach to your problem and relied upon theory that you have learned through reading, class instructions, or some past experience.

An electronics technician must use a systematic approach to every problem he encounters in faulty equipment. His resource is the previous training he has acquired, and the test of this resource is how well he can apply his knowledge in actual practice.

Let us use the radio as an example for a systematic approach to a troubleshooting problem. An analysis of this radio tells us that the input is the antenna and that the output is the speaker. The fault or trouble will lie between these two points, and your approach to this problem will be to check from one of these points. You must first look for clues that might tell you where to start. By this I mean what is the performance of the radio? Is there any type of sound coming from the speaker? If so, what is the nature of this sound? Is there a loud hum? Is the intelligence coming from the speaker distorted? If no sound is coming from the speaker, in a tube type radio, are all of the tubes lit? By visual inspection, can you see any wires that are broken or have come unsoldered? There are clues that will help you to decide where to start checking, whether at the speaker or the antenna.

In many cases the fault could be in the AC line cord or plug on an AC operated radio. Make every test possible before you decide to remove the chassis from the cabinet.

Your success or failure as an electronic technician will be your ability to understand theory and then apply that theory in a logical approach to a crisis (which is troubleshooting faulty electronic equipment). Knowing where to apply the theory you have learned is what I am stressing in this chapter.

Let us use an example of applying your theory to a certain electronic problem. Suppose you are given a piece of gear that is inoperative, and from its outward appearance you have never seen a piece of equipment like this before. This fact should not cause you to panic and give up before you start. Here is where your logical approach to the problem should be applied.

First, every piece of electronic equipment uses a combination of at least two of the three basic component parts: a resistor, a capacitor, and an inductor. These basic components, combined with other components, form an electronic circuit.

Second, you have learned from theory what these parts look like, how they are made up physically, what their purpose is in an electronic circuit, and that by combining them with other components, you have a circuit that will perform a function. Combining one or more basic circuits, you will have a complex circuit—which you are holding in your hand and saying you have never seen before.

You are given a schematic diagram of this piece of equipment. Now do you know how to read this diagram? You say, "Sure I can read it. Here is the symbol of a capacitor and there is a resistor," etc., but do you know where the starting and stopping points of this equipment are? If it is going to perform a task which it was designed for, then it must have someplace to start and someplace to stop. Now, do you know *where to start*?

It is customary, when drawing the schematic design for electronic equipment, to start on the left side of the page for the input and to show the output on the right side of the page. Also, the power supply for the equipment is usually shown in the left bottom.

This example is not much different from the radio that we used as an example. With the radio you could see the antenna and the speaker; you knew that somewhere between those two places was the trouble. With this last example you could not see the start or the output, but with the diagram you could tell that the left side of the diagram was the input and the right side was the output. Someplace between these two places is the trouble which you must localize.

LOCALIZING THE TROUBLE

If you cannot visually see the trouble, such as a loose wire, a poor solder connection, a burned resistor or one that has been overheated, maybe the equipment is one that uses a CRT for its output, such as a television receiver, an electronic typewriter, or a radar set. Here you may be able to localize the trouble by what you see or do not see on the face of the tube. Suppose the CRT does not light up; the trouble might be in the high voltage section or a defective CRT. There might be a bright horizontal line across the face of the tube; this is an indication that the trouble is in the vertical section. In this type of equipment, the CRT is telling you where the trouble might be located. If a thorough investigation does not reveal the trouble visually, then you must rely on your instruments to localize the faulty stage. You must decide what instruments to use to perform this task with the greatest accuracy and speed. You look at the diagram, and here are some things you must decide.

Is it a type of equipment to which you can apply a signal, either to the input or to the output, and trace that signal until you find the faulty stage?

Do you use an audio or RF signal for this tracing method? The oscilloscope will probably be the best instrument to use for tracing, but you will have to decide what probe to use. Remember, if you are going to check voltages, you must use a VTVM or a digital instrument.

While you are checking what instruments you are going to use, you also decide where you are going to start checking.

When you have isolated the fault to one stage, then you find the component or components that are causing the fault. You use reasoning and apply the theory you have learned. This gives you the confidence to troubleshoot any type of electronic equipment.

The requirements for a successful electronic technician in today's fast-moving society are *speed, accuracy, and results*.

REPAIR TECHNIQUES

Probably the greatest technique you should acquire at a very early state as an electronic technician is *neatness*. A satisfying moment in your electronic career is when someone says, "That surely is a neat job." Be proud of your work, and others will soon recognize your ability.

Never have more than one piece of equipment on your bench at one time. Never leave test leads and instruments leads hanging down over your bench, because when you go to use one of them, you cannot find the right one for the tangle of leads hanging from instruments. Tools that you are using at the present time should be the only ones on the test bench.

Have your work bench well lighted, preferably with an adjustable light that can be brought close to the equipment you are working on.

Use some sort of adjustable rack to hold the diagrams that you are using at the present time. This rack could be separate from the bench. This method will keep your diagrams clean and will also keep them within reach at all times. When you finish using a diagram place it back in the cabinet or file so that you will know where it is the next time you need it.

Keep your instruments in good working condition at all times. When you are not busy, go over your instruments and check them; also keep the front panels and meter face clean. You are not as likely to make a mistake in your readings if the front panel is in good visible condition.

The following are some hints you may find beneficial during your electronic career:

1. When checking a printed circuit board, if you suspect the board is cracked, a quick way to find the crack is to hold the board up between you and a strong light. The light will shine through the board, and the crack will be visible. You can use two methods for repairs. First, using a small pencil iron, heat on both sides of the crack, moving the iron back and forth across the crack. Apply a very small amount of

solder so that it runs across the crack and completes the circuit. Or you can cut a piece of small size wire, and holding it with your needle-nose pliers, solder it across the crack. This will complete the circuit.

2. When you check a transistor and find it defective, do not throw it away. A transistor is simply two diodes back to back. Use your VOM and check to see which diode is defective. Suppose the base-emitter is open or shorted, then cut the emitter off at the top of the case. Now you can use the base-collector as a signal diode. Suppose the base-collector is defective. Cut the collector off and use the base-emitter as the signal diode. You will find this method in some new equipment.

3. Ball-points pens, felt marker pens, or small pill bottles make very good substitute probe ends. An RF probe can be made by using a thin piece of fiber board. Mount on it a 4.7 meg. resistor and a .005 mfd capacitor; solder these two together along with an 1N34A (or similar diode). Solder the anode end of the diode to the resistor and capacitor. The cathode of the diode is then soldered to the shield of a piece of co-ax cable, and the resistor is soldered to the center lead of the co-ax cable. The other end of the resistor is connected to the tip, which if necessary could be a sixpenny nail.

4. If you have a small stereo amplifier, this can be used as a signal tracer when you are not using it as an amplifier. Connect the probe we have described above to one channel and make up a probe with a .01 mfd capacitor in series with the center conductor. Connect this to the other channel and use this for tracing an audio signal. If the same type of connector is used on the probe that you used for connecting your tape or record player, then it can serve two purposes.

5. When you are working on any type of equipment that does not have a transformer between the power line and the equipment, you should always use an isolation transformer and plug the equipment into the transformer. This will isolate the power line and the equipment and could prevent you from receiving a dangerous shock.

6. When using probes on test equipment, make certain that you check the voltage rating on the coupling capacitor in series with the probe. Since most of these capacitors are only rated at 600 volts, make certain that you do not use one on equipment where the voltage exceeds this rating.

Index